Nanobiotechnology: An Introduction

Nanobiotechnology: An Introduction

Edited by
Clive Jarvis

Larsen & Keller
www.larsen-keller.com

Nanobiotechnology: An Introduction
Edited by Clive Jarvis
ISBN: 978-1-63549-652-9 (Hardback)

▄ Larsen & Keller

Published by Larsen and Keller Education,
5 Penn Plaza,
19th Floor,
New York, NY 10001, USA

Cataloging-in-Publication Data

Nanobiotechnology : an introduction / edited by Clive Jarvis.
    p. cm.
Includes bibliographical references and index.
ISBN 978-1-63549-652-9
1. Nanobiotechnology. 2. Nanotechnology. 3. Biotechnology. I. Jarvis, Clive.
TP248.25.N35 N36 2018
620.5--dc23

For more information regarding Larsen and Keller Education and its products, please visit the publisher's website www.larsen-keller.com

Table of Contents

Preface

Nanobiotechnology is the integration of the principles of biology and nanotechnology. Nanobiotechnology uses equipments and materials at molecular levels with techniques of biology to understand genetic and cellular processes. It has contributed significantly towards the progress of medical science. This textbook unfolds the innovative aspects of nanobiotechnology which will be crucial for the holistic understanding of the subject matter. In this book, constant effort has been made to make the understanding of the difficult concepts of nanobiotechnology, as easy and informative as possible, for the readers.

A short introduction to every chapter is written below to provide an overview of the content of the book:

Chapter 1 - Nanobiotechnology is the culmination of nanotechnology and biology. Taking an approach to biology with the help of nanotechnology helps scientists develop and build systems and products to be used in biological researches. Nanobiotechnology is an emerging field of study; the following chapter will not only provide an overview, it will also delve deep into the variegated topics related to it; **Chapter 2** - As opposed to physical and chemical methods, green synthesis is non-toxic, cost-effective, and eco-friendly. However, it has its drawbacks. It is time-consuming and proves difficult to control size, shape and crystallinity. Nonetheless, it can optimize metal ion concentration, pH, temperature etc., and hence is widely used in biotechnology. This section discusses an important method of nanobiotechnology in a critical manner providing key analysis to the subject matter; **Chapter 3** - Nanotechnology has contributed in the development of sensors. It outweighs conventional sensors as it has higher sensitivity, consumes less power, effectively cuts production cost, and provides better stability. Biosensor, a device to detect analytes in biological systems, has gained immense popularity. Tools are an important component of any field of study. The following chapter elucidates the various tools that are related to the nanobiotechnology; **Chapter 4** - The nanoparticles that are produced through the synthesis of biological systems are called biogenic nanoparticles. As the process does not use any chemical methods, it falls under the category of green synthesis. Stealth particles are those nanoparticles that can avoid immune recognition and reach a specified target. This section has been carefully written to provide an easy understanding of the varied facets of nanobiotechnology.

I extend my sincere thanks to the publisher for considering me worthy of this task. Finally, I thank my family for being a source of support and help.

Editor

Understanding Nanobiotechnology

Nanobiotechnology is the culmination of nanotechnology and biology. Taking an approach to biology with the help of nanotechnology helps scientists develop and build systems and products to be used in biological researches. Nanobiotechnology is an emerging field of study; the following chapter will not only provide an overview, it will also delve deep into the variegated topics related to it.

Nanobiotechnology

Nanobiotechnology, bionanotechnology, and nanobiology are terms that refer to the intersection of nanotechnology and biology. Given that the subject is one that has only emerged very recently, bionanotechnology and nanobiotechnology serve as blanket terms for various related technologies.

This discipline helps to indicate the merger of biological research with various fields of nanotechnology. Concepts that are enhanced through nanobiology include: nanodevices (such as biological machines), nanoparticles, and nanoscale phenomena that occurs within the discipline of nanotechnology. This technical approach to biology allows scientists to imagine and create systems that can be used for biological research. Biologically inspired nanotechnology uses biological systems as the inspirations for technologies not yet created. However, as with nanotechnology and biotechnology, bionanotechnology does have many potential ethical issues associated with it.

The most important objectives that are frequently found in nanobiology involve applying nanotools to relevant medical/biological problems and refining these applications. Developing new tools, such as peptoid nanosheets, for medical and biological purposes is another primary objective in nanotechnology. New nanotools are often made by refining the applications of the nanotools that are already being used. The imaging of native biomolecules, biological membranes, and tissues is also a major topic for the nanobiology researchers. Other topics concerning nanobiology include the use of cantilever array sensors and the application of nanophotonics for manipulating molecular processes in living cells.

Recently, the use of microorganisms to synthesize functional nanoparticles has been of great interest. Microorganisms can change the oxidation state of metals. These microbial processes have opened up new opportunities for us to explore novel applications, for example, the biosynthesis of metal nanomaterials. In contrast to chemical and physical methods, microbial processes for synthesizing nanomaterials can be achieved in aqueous phase under gentle and environmentally benign conditions. This approach has become an attractive focus in current green bionanotechnology research towards sustainable development.

Terminology

The terms are often used interchangeably. When a distinction is intended, though, it is based on whether the focus is on applying biological ideas or on studying biology with nanotechnology. Bionanotechnology generally refers to the study of how the goals of nanotechnology can be guided by studying how biological "machines" work and adapting these biological motifs into improving existing nanotechnologies or creating new ones. Nanobiotechnology, on the other hand, refers to the ways that nanotechnology is used to create devices to study biological systems.

In other words, nanobiotechnology is essentially miniaturized biotechnology, whereas bionanotechnology is a specific application of nanotechnology. For example, DNA nanotechnology or cellular engineering would be classified as bionanotechnology because they involve working with biomolecules on the nanoscale. Conversely, many new medical technologies involving nanoparticles as delivery systems or as sensors would be examples of nanobiotechnology since they involve using nanotechnology to advance the goals of biology.

The definitions enumerated above will be utilized whenever a distinction between nanobio and bionano is made in this article. However, given the overlapping usage of the terms in modern parlance, individual technologies may need to be evaluated to determine which term is more fitting. As such, they are best discussed in parallel.

Concepts

Most of the scientific concepts in bionanotechnology are derived from other fields. Biochemical principles that are used to understand the material properties of biological systems are central in bionanotechnology because those same principles are to be used to create new technologies. Material properties and applications studied in bionanoscience include mechanical properties (e.g. deformation, adhesion, failure), electrical/electronic (e.g. electromechanical stimulation, capacitors, energy storage/batteries), optical (e.g. absorption, luminescence, photochemistry), thermal (e.g. thermomutability, thermal management), biological (e.g. how cells interact with nanomaterials, molecular flaws/defects, biosensing, biological mechanisms s.a. mechanosensing), nanoscience of disease (e.g. genetic disease, cancer, organ/tissue failure), as well as computing (e.g. DNA computing) and agriculture (target delivery of pesticides, hormones and fertilizers. The impact of bionanoscience, achieved through structural and mechanistic analyses of biological processes at nanoscale, is their translation into synthetic and technological applications through nanotechnology.

Nano-biotechnology takes most of its fundamentals from nanotechnology. Most of the devices designed for nano-biotechnological use are directly based on other existing nanotechnologies. Nano-biotechnology is often used to describe the overlapping multidisciplinary activities associated with biosensors, particularly where photonics, chemistry, biology, biophysics, nano-medicine, and engineering converge. Measurement in biology using wave guide techniques, such as dual polarization interferometry, are another example.

Applications

Applications of bionanotechnology are extremely widespread. Insofar as the distinction holds, nanobiotechnology is much more commonplace in that it simply provides more tools for the study of biology. Bionanotechnology, on the other hand, promises to recreate biological mechanisms and pathways in a form that is useful in other ways.

Nanomedicine

Nanomedicine is a field of medical science whose applications are increasing more and more thanks to nanorobots and biological machines, which constitute a very useful tool to develop this area of knowledge. In the past years, researchers have done many improvements in the different devices and systems required to develop nanorobots. This supposes a new way of treating and dealing with diseases such as cancer; thanks to nanorobots, side effects of chemotherapy have been controlled, reduced and even eliminated, so some years from now, cancer patients will be offered an alternative to treat this disease instead of chemotherapy, which causes secondary effects such as hair loss, fatigue or nausea killing not only cancerous cells but also the healthy ones. At a clinical level, cancer treatment with nanomedicine will consist on the supply of nanorobots to the patient through an injection that will seek for cancerous cells leaving untouched the healthy ones. Patients that will be treated through nanomedicine will not notice the presence of this nanomachines inside them; the only thing that is going to be noticeable is the progressive improvement of their health.

Nanobiotechnology (sometimes referred to as nanobiology) is best described as helping modern medicine progress from treating symptoms to generating cures and regenerating biological tissues. Three American patients have received whole cultured bladders with the help of doctors who use nanobiology techniques in their practice. Also, it has been demonstrated in animal studies that a uterus can be grown outside the body and then placed in the body in order to produce a baby. Stem cell treatments have been used to fix diseases that are found in the human heart and are in clinical trials in the United States. There is also funding for research into allowing people to have new limbs without having to resort to prosthesis. Artificial proteins might also become available to manufacture without the need for harsh chemicals and expensive machines. It has even been surmised that by the year 2055, computers may be made out of biochemicals and organic salts.

Another example of current nanobiotechnological research involves nanospheres coated with fluorescent polymers. Researchers are seeking to design polymers whose fluorescence is quenched when they encounter specific molecules. Different polymers would detect different metabolites. The polymer-coated spheres could become part of new biological assays, and the technology might someday lead to particles which could be introduced into the human body to track down metabolites associated with tumors and other health problems. Another example, from a different perspective, would be evaluation and therapy at the nanoscopic level, i.e. the treatment of Nanobacteria (25-200 nm sized) as is done by NanoBiotech Pharma.

While nanobiology is in its infancy, there are a lot of promising methods that will rely on nanobiology in the future. Biological systems are inherently nano in scale; nanoscience must merge with biology in order to deliver biomacromolecules and molecular machines that are similar to nature. Controlling and mimicking the devices and processes that are constructed from molecules is a tremendous challenge to face the converging disciplines of nanotechnology. All living things, including humans, can be considered to be nanofoundries. Natural evolution has optimized the "natural" form of nanobiology over millions of years. In the 21st century, humans have developed the technology to artificially tap into nanobiology. This process is best described as "organic merging with synthetic." Colonies of live neurons can live together on a biochip device; according to research from Dr. Gunther Gross at the University of North Texas. Self-assembling nanotubes have the ability to be used as a structural system. They would be composed together with rhodopsins; which would facilitate the optical computing process and help with the storage of biological materials. DNA (as the software for all living things) can be used as a structural proteomic system - a logical component for molecular computing. Ned Seeman - a researcher at New York University - along with other researchers are currently researching concepts that are similar to each other.

Bionanotechnology

DNA nanotechnology is one important example of bionanotechnology. The utilization of the inherent properties of nucleic acids like DNA to create useful materials is a promising area of modern research. Another important area of research involves taking advantage of membrane properties to generate synthetic membranes. Proteins that self-assemble to generate functional materials could be used as a novel approach for the large-scale production of programmable nanomaterials. One example is the development of amyloids found in bacterial biofilms as engineered nanomaterials that can be programmed genetically to have different properties. Protein folding studies provide a third important avenue of research, but one that has been largely inhibited by our inability to predict protein folding with a sufficiently high degree of accuracy. Given the myriad uses that biological systems have for proteins, though, research into understanding protein folding is of high importance and could prove fruitful for bionanotechnology in the future.

Lipid nanotechnology is another major area of research in bionanotechnology, where physico-chemical properties of lipids such as their antifouling and self-assembly is exploited to build nanodevices with applications in medicine and engineering.

Agriculture

Meanwhile, nanotechnology application to biotechnology will also leave no field untouched by its groundbreaking scientific innovations for human wellness; the agricultural industry is no exception. Basically, nanomaterials are distinguished depending on the origin: natural, incidental and engineered nanoparticles. Among these, engineered nanoparticles have received wide attention in all fields of science, including medical, materials and agriculture technology with significant socio-economical growth. In the agriculture industry, engineered nanoparticles have been serving as nano carrier, containing herbicides, chemicals, or genes, which target particular plant parts to release their content. Previously nanocapsules containing herbicides have been reported to effectively penetrate through cuticles and tissues, allowing the slow and constant release of the active substances. Likewise, other literature describes that nano-encapsulated slow release of fertilizers has also become a trend to save fertilizer consumption and to minimize environmental pollution through precision farming. These are only a few examples from numerous research works which might open up exciting opportunities for nanobiotechnology application in agriculture. Also, application of this kind of engineered nanoparticles to plants should be considered the level of amicability before it is employed in agriculture practices. Based on a thorough literature survey, it was understood that there is only limited authentic information available to explain the biological consequence of engineered nanoparticles on treated plants. Certain reports underline the phytotoxicity of various origin of engineered nanoparticles to the plant caused by the subject of concentrations and sizes . At the same time, however, an equal number of studies were reported with a positive outcome of nanoparticles, which facilitate growth promoting nature to treat plant. In particular, compared to other nanoparticles, silver and gold nanoparticles based applications elicited beneficial results on various plant species with less and/or no toxicity. Silver nanoparticles (AgNPs) treated leaves of Asparagus showed the increased content of ascorbate and chlorophyll. Similarly, AgNPs-treated common bean and corn has increased shoot and root length, leaf surface area, chlorophyll, carbohydrate and protein contents reported earlier. The gold nanoparticle has been used to induce growth and seed yield in Brassica juncea.

Tools

This field relies on a variety of research methods, including experimental tools (e.g. imaging, characterization via AFM/optical tweezers etc.), x-ray diffraction based tools, synthesis via self-assembly, characterization of self-assembly (using e.g. MP-SPR, DPI, recombinant DNA methods, etc.), theory (e.g. statistical mechanics, nanomechanics, etc.), as well as computational approaches (bottom-up multi-scale simulation, supercomputing).

Nanorods

The shape of the nano sized particles plays a key role in stratifying the purpose of the nano particle. Up till now, out of the many developed different shapes, Nanorods are among the most widely used one. Just as the name suggests, these nano rods are the elongated shaped, nano sized objects which have proved to be useful in many technological amelioration. These are generally manufactured using metals and semiconductors. The easiest and the most cost effective manufacturing is through direct chemical synthesis using specific chemical reagents which act as ligand during the formation and by their capability to combine specifically, they are the shape governing factor for nanorods.

The feature that makes nanorods so special is their anisotropic shape i.e. due to their shape, various properties like refractive index, adsorption etc. differ in different direction. Due to availability of two direction in Rods (transverse direction and longitudinal direction), this phenomenon of change in properties can be used very effectively in Biosensing, Imaging and various other important phenomena. The change in the outer environment leads to aligning of these particles in different ways which leads to change in the wavelength they emit, thus acing as a good sensor which changes the color as soon as there is any change in the environment.

Preparation of Gold Nanorods

As stated above, the best method to synthesize Gold Nanorods is the direct chemical method. The process involves the introduction of Gold Nano particles (spherical in shape) in a growth solution containing Gold salt, Silver Nitrate, Cetyltrimethylammonium bromide. The Cetyltrimethyl ammonium bromide acts as the growth directing agent . T binds preferentially to one side of the spherical gold anno particles and thus help in formation of Gold Nanorods. Once , the process if completed, excess of Cetyltrimethyl ammonium bromide is removed by centrifugation because of the toxicity associated with it. The other unreacted species are also removed because they may lead to change in morphology over a prolonged period.

Various Applications of Gold Nanorods :

1. Bio sensing

2. Drug Delivery

3. Tracking and Imaging

4. Photothermal Therapeutics

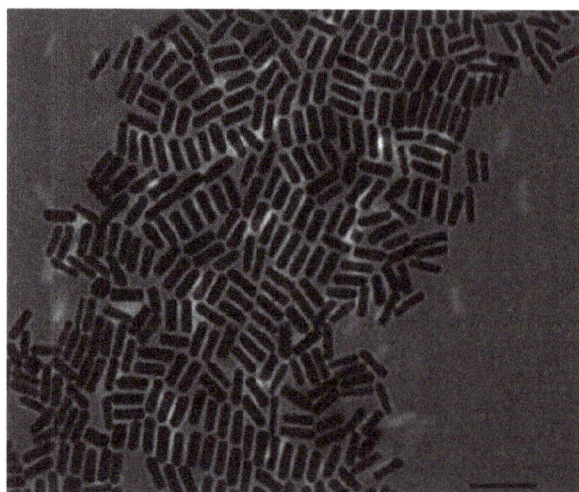

Chemical Synthesis

Chemical synthesis is the purposeful execution of one or more named reactions to obtain a product, or several products. In modern laboratory usage, this tends to imply that the process is reproducible, reliable, and established to work in multiple laboratories.

Introduction

A chemical synthesis—synthesis, in its present meaning, begins with the selection of chemical product target, which often possesses academic, industrial or therapeutic interest to the broader aims of the research effort. Secondary practical concerns also come into play, including researcher availability, availability of material resources (necessary equipment, chemical building blocks), and research budget. A research grant proposal is often submitted to a funding agency (such as National Science Foundation or National Institutes of Health), which describes the background/proposed synthesis and research plans, and secondary concerns. For instance, a prior developed reaction methodology may highlight particular man-made or natural compounds that would serve the purposes of the effort, in highlighting the breadth of the new methodology and possibly providing facile access to complex natural products. Once the target or targets are established, the next critical phase begins, that of synthetic design, typically in modern efforts, in the area of organic synthesis, using retrosynthetic analysis, as championed by E.J. Corey and others.

An eventual step is a selection of compounds that are known as reagents or reactants. Reactants are compounds used in a reaction that combines to form the product of the reaction. Various reaction types can be applied to these to synthesize the product or an intermediate product. This requires mixing the compounds in a reaction vessel such as a chemical reactor or a simple round-bottom flask. Many reactions require some form of work-up procedure before the final product is isolated. The isolation (purification) of the product then proceeds via a variety of methods.

The amount of product in a chemical synthesis is the reaction yield. Typically, chemical yields are expressed as a weight in grams (in a laboratory setting) or as a percentage of the total theoretical quantity of product that could be produced. A side reaction is an unwanted chemical reaction taking place that diminishes the yield of the desired product.

Strategies

Many strategies exist in chemical synthesis that go beyond converting reactant A to reaction product B in a single step. In multistep synthesis, a chemical compound is synthesized through a series of individual chemical reactions, each with their own workup. For example, a laboratory synthesis of paracetamol can consist of three individual synthetic steps. In cascade reactions multiple chemical transformations take place within a single reactant, in multi-component reactions up to 11 different reactants form a single reaction product and in a telescopic synthesis one reactant goes through multiple transformations without isolation of intermediates.

Organic Synthesis

Organic synthesis is a special branch of chemical synthesis dealing with the synthesis of organic compounds. In the total synthesis of a complex product it may take multiple steps to synthesize the product of interest, and inordinate amounts of time. Skill in organic synthesis is prized among chemists and the synthesis of exceptionally valuable or difficult compounds has won chemists such as Robert Burns Woodward the Nobel Prize for Chemistry. If a chemical synthesis starts from basic laboratory compounds and yields something new, it is a purely synthetic process. If it starts from a product isolated from plants or animals and then proceeds to new compounds, the synthesis is described as a semisynthetic process.

Ligand

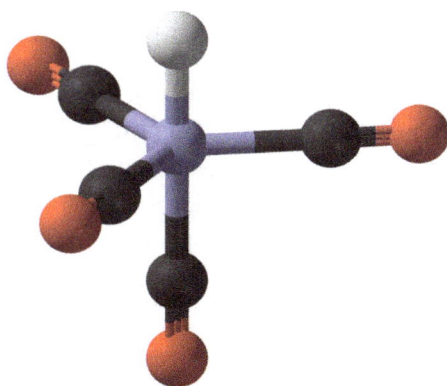

Cobalt complex $HCo(CO)_4$ with five ligands

In coordination chemistry, a ligand is an ion or molecule (functional group) that binds to a central metal atom to form a coordination complex. The bonding with the metal generally involves formal donation of one or more of the ligand's electron pairs. The nature of metal–ligand bonding can range from covalent to ionic. Furthermore, the metal–ligand bond order can range from one to three. Ligands are viewed as Lewis bases, although rare cases are known to involve Lewis acidic "ligand".

Metals and metalloids are bound to ligands in virtually all circumstances, although gaseous "naked" metal ions can be generated in high vacuum. Ligands in a complex dictate the reactivity of the central atom, including ligand substitution rates, the reactivity of the ligands themselves, and redox. Ligand selection is a critical consideration in many practical areas, including bioinorganic and medicinal chemistry, homogeneous catalysis, and environmental chemistry.

Ligands are classified in many ways, including: charge, size (bulk), the identity of the coordinating atom(s), and the number of electrons donated to the metal (denticity or hapticity). The size of a ligand is indicated by its cone angle.

History

The composition of coordination complexes have been known since the early 1800s, such as Prussian blue and copper vitriol. The key breakthrough occurred when Alfred Werner reconciled formulas and isomers. He showed, among other things, that the formulas of many cobalt(III) and chromium(III) compounds can be understood if the metal has six ligands in an octahedral geometry. The first to use the term "ligand" were Alfred Stock and Carl Somiesky, in relation to silicon chemistry. The theory allows one to understand the difference between coordinated and ionic chloride in the cobalt ammine chlorides and to explain many of the previously inexplicable isomers. He resolved the first coordination complex called hexol into optical isomers, overthrowing the theory that chirality was necessarily associated with carbon compounds.

Strong Field and Weak Field Ligands

In general, ligands are viewed as electron donors and the metals as electron acceptors. This is because the ligand and central metal are bonded to one another, and the ligand is providing both electrons to the bond (lone pair of electrons) instead of the metal and ligand each providing one electron. Bonding is often described using the formalisms of molecular orbital theory. The HOMO (Highest Occupied Molecular Orbital) can be mainly of ligands or metal character.

Ligands and metal ions can be ordered in many ways; one ranking system focuses on ligand 'hardness'. Metal ions preferentially bind certain ligands. In general, 'soft' metal ions prefer weak field ligands, whereas 'hard' metal ions prefer strong field ligands. According to the molecular orbital theory, the HOMO of the ligand should have an energy that overlaps with the LUMO (Lowest Unoccupied Molecular Orbital) of the metal preferential. Metal ions bound to strong-field ligands follow the Aufbau principle, whereas complexes bound to weak-field ligands follow Hund's rule.

Binding of the metal with the ligands results in a set of molecular orbitals, where the metal can be identified with a new HOMO and LUMO (the orbitals defining the properties and reactivity of the resulting complex) and a certain ordering of the 5 d-orbitals (which may be filled, or partially filled with electrons). In an octahedral environment, the 5 otherwise degenerate d-orbitals split in sets of 2 and 3 orbitals.

3 orbitals of low energy: d_{xy}, d_{xz} and d_{yz}

2 of high energy: d_{z^2} and $d_{x^2-y^2}$

The energy difference between these 2 sets of d-orbitals is called the splitting parameter, Δ_o. The magnitude of Δ_o is determined by the field-strength of the ligand: strong field ligands, by definition, increase Δ_o more than weak field ligands. Ligands can now be sorted according to the magnitude of Δ_o. This ordering of ligands is almost invariable for all metal ions and is called spectrochemical series.

For complexes with a tetrahedral surrounding, the d-orbitals again split into two sets, but this time in reverse order.

2 orbitals of low energy: d_{z^2} and $d_{x^2-y^2}$

3 orbitals of high energy: d_{xy}, d_{xz} and d_{yz}

The energy difference between these 2 sets of d-orbitals is now called Δ_t. The magnitude of Δ_t is smaller than for Δ_o, because in a tetrahedral complex only 4 ligands influence the d-orbitals, whereas in an octahedral complex the d-orbitals are influenced by 6 ligands. When the coordination number is neither octahedral nor tetrahedral, the splitting becomes correspondingly more complex. For the purposes of ranking ligands, however, the properties of the octahedral complexes and the resulting Δ_o has been of primary interest.

The arrangement of the d-orbitals on the central atom (as determined by the 'strength' of the ligand), has a strong effect on virtually all the properties of the resulting complexes. E.g., the energy differences in the d-orbitals has a strong effect in the optical absorption spectra of metal complexes. It turns out that valence electrons occupying orbitals with significant 3 d-orbital character absorb in the 400–800 nm region of the spectrum (UV–visible range). The absorption of light (what we perceive as the color) by these electrons (that is, excitation of electrons from one orbital to another orbital under influence of light) can be correlated to the ground state of the metal complex, which reflects the bonding properties of the ligands. The relative change in (relative) energy of the d-orbitals as a function of the field-strength of the ligands is described in Tanabe–Sugano diagrams.

In cases where the ligand has low energy LUMO, such orbitals also participate in the bonding. The metal–ligand bond can be further stabilised by a formal donation of electron density back to the ligand in a process known as *back-bonding*. In this case a filled, central-atom-based orbital donates density into the LUMO of the (coordinated) ligand. Carbon monoxide is the preeminent example a ligand that engages metals via back-donation. Complementarily, ligands with low-energy filled orbitals of pi-symmetry can serve as pi-donor.

Classification of Ligands as L and X

Especially in the area of organometallic chemistry, ligands are classified as L and X (or combinations of the two). The classification scheme – the "CBC Method" for Covalent Bond Classification – was popularized by M.L.H. Green and "is based on the notion that there are three basic types [of ligands]… represented by the symbols L, X, and Z, which correspond respectively to 2-electron, 1-electron and 0-electron neutral ligands." Another type of ligand worthy of consideration is the LX ligand which as expected from the used conventional representation will donate three electrons if NVE (Number of Valence Electrons) required. Example is alkoxy ligands(which is regularly known as X ligand too). L ligands are derived from charge-neutral precursors and are represented by amines, phosphines, CO, N_2, and alkenes. X ligands typically are derived from anionic precursors such as chloride but includes ligands where salts of anion do not really exist such as hydride and alkyl. Thus, the complex $IrCl(CO)$ $(PPh_3)_2$ is classified as an MXL_3 complex, since CO and the two PPh_3 ligands are classified as Ls. The oxidative addition of H_2 to $IrCl(CO)(PPh_3)_2$ gives an $18e^-$ ML_3X_3 product, $IrClH_2(CO)$ $(PPh_3)_2$. $EDTA^{4-}$ is classified as an L_2X_4 ligand, as it features four anions and two neutral donor sites. Cp is classified as an L_2X ligand.

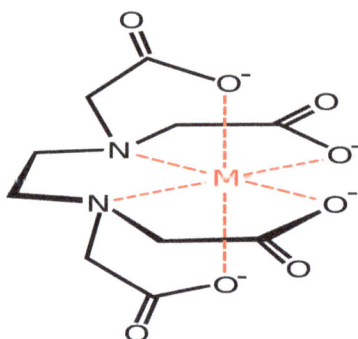

Metal–EDTA complex, wherein the aminocarboxylate is a hexadentate (chelating ligand).

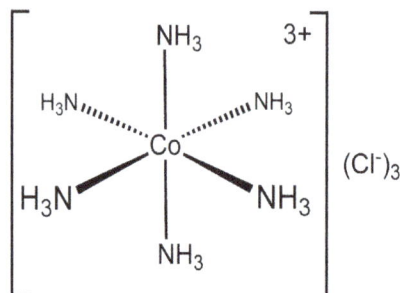

Cobalt(III) complex containing six ammonia ligands, which are monodentate. The chloride is not a ligand.

Polydentate and Polyhapto Ligand Motifs and Nomenclature

Denticity

Denticity (represented by κ) refers to the number of times a ligand bonds to a metal through non-contiguous donor sites. Many ligands are capable of binding metal ions through multiple sites, usually because the ligands have lone pairs on more than one atom. Ligands that bind via more than one atom are often termed *chelating*. A ligand that binds through two sites is classified as *bidentate*, and three sites as *tridentate*. The "bite angle" refers to the angle between the two bonds of a bidentate chelate. Chelating ligands are commonly formed by linking donor groups via organic linkers. A classic bidentate ligand is ethylenediamine, which is derived by the linking of two ammonia groups with an ethylene ($-CH_2CH_2-$) linker. A classic example of a polydentate ligand is the hexadentate chelating agent EDTA, which is able to bond through six sites, completely surrounding some metals. The number of times a polydentate ligand binds to a metal centre is symbolized by "κ^n", where n indicates the number of sites by which a ligand attaches to a metal. $EDTA^{4-}$, when it is hexidentate, binds as a κ^6-ligand, the amines and the carboxylate oxygen atoms are not contiguous. In practice, the n value of a ligand is not indicated explicitly but rather assumed. The binding affinity of a chelating system depends on the chelating angle or bite angle.

Complexes of polydentate ligands are called *chelate* complexes. They tend to be more stable than complexes derived from monodentate ligands. This enhanced stability, the chelate effect, is usually attributed to effects of entropy, which favors the displacement of many ligands by one polydentate ligand. When the chelating ligand forms a large ring that at least partially surrounds the central atom and bonds to it, leaving the central atom at the centre of a large ring. The more rigid and the higher its denticity, the more inert will be the macrocyclic complex. Heme is a good example: the iron atom is at the centre of a porphyrin macrocycle, being bound to four nitrogen atoms of the tetrapyrrole macrocycle. The very stable dimethylglyoximate complex of nickel is a synthetic macrocycle derived from the anion of dimethylglyoxime.

Hapticity

Hapticity (represented by η) refers to the number of *contiguous* atoms that comprise a donor site and attach to a metal center. Butadiene forms both η^2 and η^4 complexes depending on the number of carbon atoms that are bonded to the metal.

Ligand Motifs

Trans-spanning Ligands

Trans-spanning ligands are bidentate ligands that can span coordination positions on opposite sides of a coordination complex.

Ambidentate Ligand

Unlike polydentate ligands, ambidentate ligands can attach to the central atom in two places. A good example of this is thiocyanate, SCN^-, which can attach at either the sulfur atom or the nitrogen atom. Such compounds give rise to linkage isomerism. Polyfunctional ligands, see especially proteins, can bond to a metal center through different ligand atoms to form various isomers.

Bridging Ligand

A bridging ligand links two or more metal centers. Virtually all inorganic solids with simple formulas are coordination polymers, consisting of metal centres linked by bridging ligands. This group of materials includes all anhydrous binary metal halides and pseudohalides. Bridging ligands also persist in solution. Polyatomic ligands such as carbonate are ambidentate and thus are found to often bind to two or three metals simultaneously. Atoms that bridge metals are sometimes indicated with the prefix "μ". Most inorganic solids are polymers by virtue of the presence of multiple bridging ligands.

Binucleating Ligand

Binucleating ligands bind two metals. Usually binucleating ligands feature bridging ligands, such as phenoxide, pyrazolate, or pyrazine, as well as other donor groups that bind to only one of the two metals.

Metal–ligand Multiple Bond

Metal ligand multiple bonds some ligands can bond to a metal center through the same atom but with a different number of lone pairs. The bond order of the metal ligand bond can be in part distinguished through the metal ligand bond angle (M–X–R). This bond angle is often referred to as being linear or bent with further discussion concerning the degree to which the angle is bent. For example, an imido ligand in the ionic form has three lone pairs. One lone pair is used as a sigma X donor, the other two lone pairs are available as L-type pi donors. If both lone pairs are used in pi bonds then the M–N–R geometry is linear. However, if one or both these lone pairs is nonbonding then the M–N–R bond is bent and the extent of the bend speaks to how much pi bonding there may be. η^1-Nitric oxide can coordinate to a metal center in linear or bent manner.

Spectator Ligand

A spectator ligand is a tightly coordinating polydentate ligand that does not participate in chemical reactions but removes active sites on a metal. Spectator ligands influence the reactivity of the metal center to which they are bound.

Bulky Ligands

Bulky ligands are used to control the steric properties of a metal center. They are used for many reasons, both practical and academic. On the practical side, they influence the selectivity of metal catalysts, e.g., in hydroformylation. Of academic interest, bulky ligands stabilize unusual coordination sites, e.g., reactive coligands or low coordination numbers. Often bulky ligands are employed to simulate the steric protection afforded by proteins to metal-containing active sites. Of course excessive steric bulk can prevent the coordination of certain ligands.

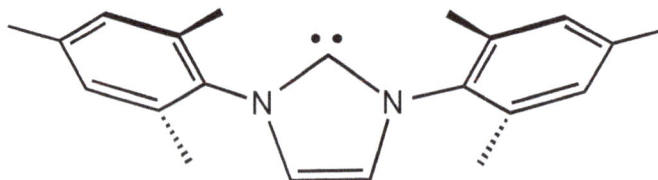

The *N*-heterocyclic carbene ligand called IMes is a bulky ligand by virtue of the pair of mesityl groups.

Chiral Ligands

Chiral ligands are useful for inducing asymmetry within the coordination sphere. Often the ligand is employed as an optically pure group. In some cases, such as secondary amines, the asymmetry arises upon coordination. Chiral ligands are used in homogeneous catalysis, such as asymmetric hydrogenation.

Hemilabile Ligands

Hemilabile ligands contain at least two electronically different coordinating groups and form complexes where one these is easily displaced from the metal center while the other remains firmly bound; a behaviour which has been found to increase the reactivity of catalysts when compared to the use of more traditional ligands.

Non-innocent Ligand

Non-innocent ligands bond with metals in such a manner that the distribution of electron density between the metal center and ligand is unclear. Describing the bonding of non-innocent ligands often involves writing multiple resonance forms that have partial contributions to the overall state.

Common Ligands

Virtually every molecule and every ion can serve as a ligand for (or "coordinate to") metals. Monodentate ligands include virtually all anions and all simple Lewis bases. Thus, the halides and pseudohalides are important anionic ligands whereas ammonia, carbon monoxide, and water are particularly common charge-neutral ligands. Simple organic species are also very common, be they anionic (RO^- and RCO_2^-) or neutral (R_2O, R_2S, $R_{3-x}NH_x$, and R_3P). The steric properties of some ligands are evaluated in terms of their cone angles.

Beyond the classical Lewis bases and anions, all unsaturated molecules are also ligands, utilizing their pi electrons in forming the coordinate bond. Also, metals can bind to the σ bonds in for example silanes, hydrocarbons, and dihydrogen.

In complexes of non-innocent ligands, the ligand is bonded to metals via conventional bonds, but the ligand is also redox-active.

Carbon Nanotube

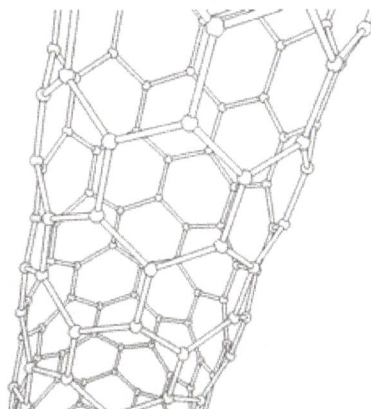

Rotating single-walled zigzag carbon nanotube

Carbon nanotubes (CNTs) are allotropes of carbon with a cylindrical nanostructure. These cylindrical carbon molecules have unusual properties, which are valuable for nanotechnology, electronics, optics and other fields of materials science and technology. Owing to the material's exceptional strength and stiffness, nanotubes have been constructed with length-to-diameter ratio of up to 132,000,000:1, significantly larger than for any other material.

In addition, owing to their extraordinary thermal conductivity, mechanical, and electrical properties, carbon nanotubes find applications as additives to various structural materials. For instance, nanotubes form a tiny portion of the material(s) in some (primarily carbon fiber) baseball bats, golf clubs, car parts or damascus steel.

Nanotubes are members of the fullerene structural family. Their name is derived from their long, hollow structure with the walls formed by one-atom-thick sheets of carbon, called graphene. These sheets are rolled at specific and discrete ("chiral") angles, and the combination of the rolling angle and radius decides the nanotube properties; for example, whether the individual nanotube shell is a metal or semiconductor. Nanotubes are categorized as single-walled nanotubes (SWNTs) and multi-walled nanotubes (MWNTs). Individual nanotubes naturally align themselves into "ropes" held together by van der Waals forces, more specifically, pi-stacking.

Applied quantum chemistry, specifically, orbital hybridization best describes chemical bonding in nanotubes. The chemical bonding of nanotubes is composed entirely of sp^2 bonds, similar to those of graphite. These bonds, which are stronger than the sp^3 bonds found in alkanes and diamond, provide nanotubes with their unique strength.

Types of Carbon Nanotubes and Related Structures

There is no consensus on some terms describing carbon nanotubes in scientific literature: both "-wall" and "-walled" are being used in combination with "single", "double", "triple" or "multi", and the letter C is often omitted in the abbreviation; for example, multi-walled carbon nanotube (MWNT).

Single-walled

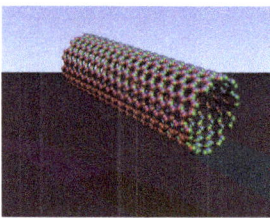 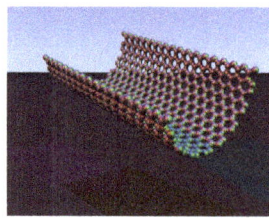 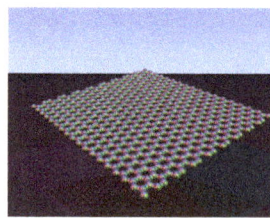 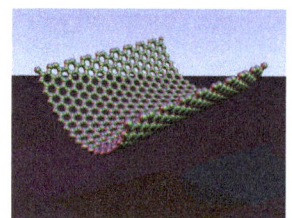

Armchair (n,n) i.e.: m=n	The translation vector is bent, while the chiral vector stays straight	Graphene nanoribbon	The chiral vector is bent, while the translation vector stays straight

 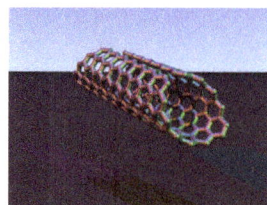 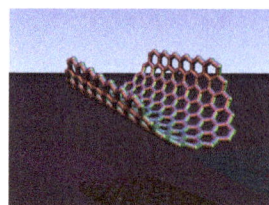 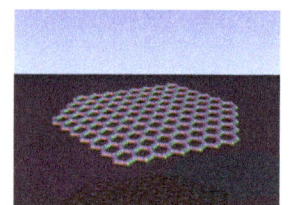

Zigzag $(n,0)$	Chiral (n,m)	n and m can be counted at the end of the tube	Graphene nanoribbon

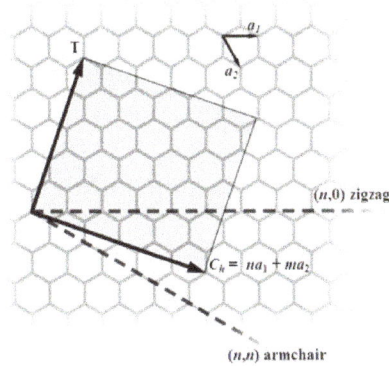

The (n,m) nanotube naming scheme can be thought of as a vector (C_h) in an infinite graphene sheet that describes how to "roll up" the graphene sheet to make the nanotube. T denotes the tube axis, and a_1 and a_2 are the unit vectors of graphene in real space.

A scanning tunneling microscopy image of single-walled carbon nanotube

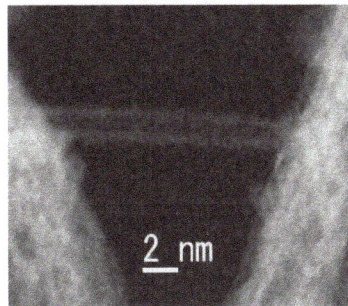

A transmission electron microscopy image of a single-walled carbon nanotube

Most single-walled nanotubes (SWNTs) have a diameter of close to 1 nanometer, and can be many millions of times longer. The structure of a SWNT can be conceptualized by wrapping a one-atom-thick layer of graphite called graphene into a seamless cylinder. The way the graphene sheet is wrapped is represented by a pair of indices (n,m). The integers n and m denote the number of unit vectors along two directions in the honeycomb crystal lattice of graphene. If m = 0, the nanotubes are called zigzag nanotubes, and if n = m, the nanotubes are called armchair nanotubes. Otherwise, they are called chiral. The diameter of an ideal nanotube can be calculated from its (n,m) indices as follows

$$d = \frac{a}{\pi}\sqrt{(n^2 + nm + m^2)} = 78.3\sqrt{((n+m)^2 - nm)}\,\text{pm},$$

where a = 0.246 nm.

SWNTs are an important variety of carbon nanotube because most of their properties change significantly with the (n,m) values, and this dependence is non-monotonic. In particular, their band gap can vary from zero to about 2 eV and their electrical conductivity can show metallic or semi-conducting behavior. Single-walled nanotubes are likely candidates for miniaturizing electronics. The most basic building block of these systems is the electric wire, and SWNTs with diameters of an order of a nanometer can be excellent conductors. One useful application of SWNTs is in the development of the first intermolecular field-effect transistors (FET). The first intermolecular logic gate using SWCNT FETs was made in 2001. A logic gate requires both a p-FET and an n-FET. Because SWNTs are p-FETs when exposed to oxygen and n-FETs otherwise, it is possible to expose half of an SWNT to oxygen and protect the other half from it. The resulting SWNT acts as a not logic gate with both p and n-type FETs in the same molecule.

Prices for single-walled nanotubes declined from around $1500 per gram as of 2000 to retail prices of around $50 per gram of as-produced 40–60% by weight SWNTs as of March 2010. As of 2016 the retail price of as-produced 75% by weight SWNTs were $2 per gram, cheap enough for widespread use. SWNTs are forecast to make a large impact in electronics applications by 2020 according to the The Global Market for Carbon Nanotubes report.

Multi-walled

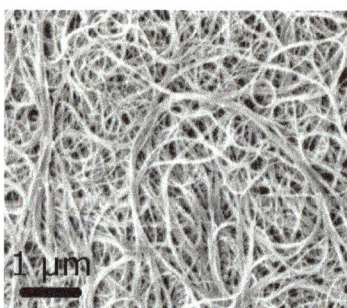

A scanning electron microscopy image of carbon nanotubes bundles

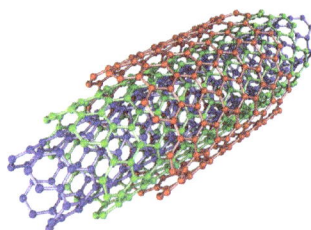

Triple-walled Armchair Carbon Nanotube

Multi-walled nanotubes (MWNTs) consist of multiple rolled layers (concentric tubes) of graphene. There are two models that can be used to describe the structures of multi-walled nanotubes. In the Russian Doll model, sheets of graphite are arranged in concentric cylinders, e.g., a (0,8) single-walled nanotube (SWNT) within a larger (0,17) single-walled nanotube. In the Parchment model, a single sheet of graphite is rolled in around itself, resembling a scroll of parchment or a rolled newspaper. The interlayer distance in multi-walled nanotubes is close to the distance between graphene layers in graphite, approximately 3.4 Å. The Russian Doll structure is observed more commonly. Its individual shells can be described as SWNTs, which can be metallic or semiconducting. Because of statistical probability and restrictions on the relative diameters of the individual tubes, one of the shells, and thus the whole MWNT, is usually a zero-gap metal.

Double-walled carbon nanotubes (DWNTs) form a special class of nanotubes because their morphology and properties are similar to those of SWNTs but they are more resistant to chemicals. This is especially important when it is necessary to graft chemical functions to the surface of the nanotubes (functionalization) to add properties to the CNT. Covalent functionalization of SWNTs will break some C=C double bonds, leaving "holes" in the structure on the nanotube, and thus modifying both its mechanical and electrical properties. In the case of DWNTs, only the outer wall is modified. DWNT synthesis on the gram-scale was first proposed in 2003 by the CCVD technique, from the selective reduction of oxide solutions in methane and hydrogen.

The telescopic motion ability of inner shells and their unique mechanical properties will permit the use of multi-walled nanotubes as main movable arms in coming nanomechanical devices. Retraction force that occurs to telescopic motion caused by the Lennard-Jones interaction between shells and its value is about 1.5 nN.

Junctions and Crosslinking

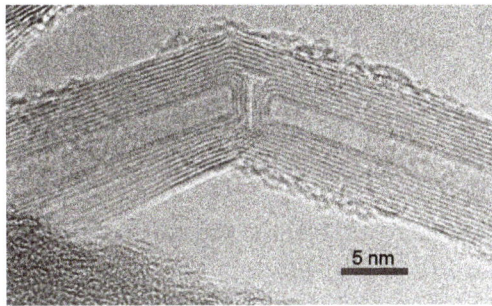

Transmission electron microscope image of carbon nanotube junction

Junctions between 2 or more nanotubes have been widely discussed theoretically. Such junctions are quite frequently observed in samples prepared by arc discharge as well as by chemical vapor deposition. The electronic properties of such junctions were first considered theoretically by Lambin et al., who pointed out that a connection between metallic tube and a semiconducting one would represent a nanoscale heterojunction. Such a junction could therefore form a component of a nanotube-based electronic circuit. The image on the right shows a junction between two multi-walled nanotubes. Junctions between nanotubes and graphene have been considered theoretically, but not widely studied experimentally. Such junctions form the basis of pillared graphene, in which parallel graphene sheets are separated by short nanotubes. Pillared graphene represents a class of three-dimensional carbon nanotube architectures.

3D carbon scaffolds

Recently, several studies have highlighted the prospect of using carbon nanotubes as building blocks to fabricate three-dimensional macroscopic (>100 nm in all three dimensions) all-carbon devices. Lalwani et al. have reported a novel radical initiated thermal crosslinking method to fabricate macroscopic, free-standing, porous, all-carbon scaffolds using single- and multi-walled carbon nanotubes as building blocks. These scaffolds possess macro-, micro-, and nano- structured pores and the porosity can be tailored for specific applications. These 3D all-carbon scaffolds/architectures may be used for the fabrication of the next generation of energy storage, supercapacitors, field emission transistors, high-performance catalysis, photovoltaics, and biomedical devices and implants.

Other Morphologies

A stable nanobud structure

Carbon nanobuds are a newly created material combining two previously discovered allotropes of carbon: carbon nanotubes and fullerenes. In this new material, fullerene-like "buds" are covalently bonded to the outer sidewalls of the underlying carbon nanotube. This hybrid material has useful properties of both fullerenes and carbon nanotubes. In particular, they have been found to be exceptionally good field emitters. In composite materials, the attached fullerene molecules may function as molecular anchors preventing slipping of the nanotubes, thus improving the composite's mechanical properties.

A carbon peapod is a novel hybrid carbon material which traps fullerene inside a carbon nanotube. It can possess interesting magnetic properties with heating and irradiation. It can also be applied as an oscillator during theoretical investigations and predictions.

In theory, a nanotorus is a carbon nanotube bent into a torus (doughnut shape). Nanotori are predicted to have many unique properties, such as magnetic moments 1000 times larger than previously expected for certain specific radii. Properties such as magnetic moment, thermal stability, etc. vary widely depending on radius of the torus and radius of the tube.

Graphenated carbon nanotubes are a relatively new hybrid that combines graphitic foliates grown along the sidewalls of multiwalled or bamboo style CNTs. The foliate density can vary as a function of deposition conditions (e.g. temperature and time) with their structure ranging from few layers of graphene (< 10) to thicker, more graphite-like. The fundamental advantage of an integrated graphene-CNT structure is the high surface area three-dimensional framework of the CNTs coupled with the high edge density of graphene. Depositing a high density of graphene foliates along the length of aligned CNTs can significantly increase the total charge capacity per unit of nominal area as compared to other carbon nanostructures.

Cup-stacked carbon nanotubes (CSCNTs) differ from other quasi-1D carbon structures, which normally behave as quasi-metallic conductors of electrons. CSCNTs exhibit semiconducting behaviors due to the stacking microstructure of graphene layers.

Extreme Carbon Nanotubes

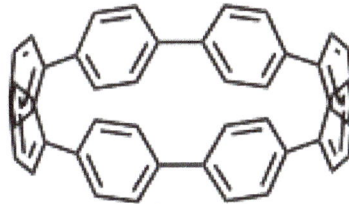

Cycloparaphenylene

The observation of the longest carbon nanotubes grown so far are over 1/2 m (550 mm long) was reported in 2013. These nanotubes were grown on Si substrates using an improved chemical vapor deposition (CVD) method and represent electrically uniform arrays of single-walled carbon nanotubes.

The shortest carbon nanotube is the organic compound cycloparaphenylene, which was synthesized in 2008.

The thinnest carbon nanotube is the armchair (2,2) CNT with a diameter of 0.3 nm. This nanotube was grown inside a multi-walled carbon nanotube. Assigning of carbon nanotube type was done by a combination of high-resolution transmission electron microscopy (HRTEM), Raman spectroscopy and density functional theory (DFT) calculations.

The thinnest freestanding single-walled carbon nanotube is about 0.43 nm in diameter. Researchers suggested that it can be either (5,1) or (4,2) SWCNT, but the exact type of carbon nanotube remains questionable. (3,3), (4,3) and (5,1) carbon nanotubes (all about 0.4 nm in diameter) were unambiguously identified using aberration-corrected high-resolution transmission electron microscopy inside double-walled CNTs.

The highest density of CNTs was achieved in 2013, grown on a conductive titanium-coated copper surface that was coated with co-catalysts cobalt and molybdenum at lower than typical temperatures of 450 °C. The tubes averaged a height of 380 nm and a mass density of 1.6 g cm^{-3}. The material showed ohmic conductivity (lowest resistance ~22 kΩ).

Properties

Mechanical

Carbon nanotubes are the strongest and stiffest materials yet discovered in terms of tensile strength and elastic modulus respectively. This strength results from the covalent sp^2 bonds formed between the individual carbon atoms. In 2000, a multi-walled carbon nanotube was tested to have a tensile strength of 63 gigapascals (9,100,000 psi). (For illustration, this translates into the ability to endure tension of a weight equivalent to 6,422 kilograms-force (62,980 N; 14,160 lbf) on a cable with cross-section of 1 square millimetre (0.0016 sq in).) Further studies, such as one conducted in 2008, revealed that individual CNT shells have strengths of up to ≈100 gigapascals (15,000,000 psi), which is in agreement with quantum/atomistic models. Since carbon nanotubes have a low density for a solid of 1.3 to 1.4 g/cm^3, its specific strength of up to 48,000 kN·m·kg^{-1} is the best of known materials, compared to high-carbon steel's 154 kN·m·kg^{-1}.

Although the strength of individual CNT shells is extremely high, weak shear interactions between adjacent shells and tubes lead to significant reduction in the effective strength of multi-walled carbon nanotubes and carbon nanotube bundles down to only a few GPa. This limitation has been recently addressed by applying high-energy electron irradiation, which crosslinks inner shells and tubes, and effectively increases the strength of these materials to ≈60 GPa for multi-walled carbon nanotubes and ≈17 GPa for double-walled carbon nanotube bundles. CNTs are not nearly as strong under compression. Because of their hollow structure and high aspect ratio, they tend to undergo buckling when placed under compressive, torsional, or bending stress.

On the other hand, there was evidence that in the radial direction they are rather soft. The first transmission electron microscope observation of radial elasticity suggested that even the van der Waals forces can deform two adjacent nanotubes. Later, nanoindentations with atomic force microscope were performed by several groups to quantitatively measure radial elasticity of multi-walled carbon nanotubes and tapping/contact mode atomic force microscopy was also performed on single-walled carbon nanotubes. Young's modulus of on the order of several GPa showed that CNTs are in fact very soft in the radial direction.

Electrical

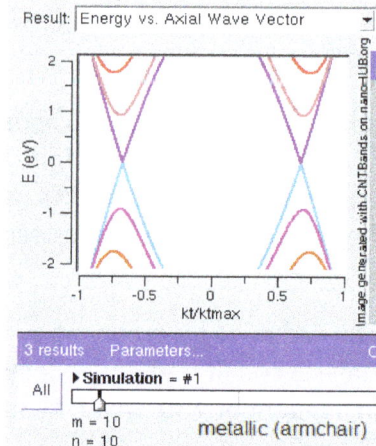

Band structures computed using tight binding approximation for (6,0) CNT (zigzag, metallic), (10,2) CNT (semiconducting) and (10,10) CNT (armchair, metallic).

Unlike graphene, which is a two-dimensional semimetal, carbon nanotubes are either metallic or semiconducting along the tubular axis. For a given (n,m) nanotube, if n = m, the nanotube is metallic; if n − m is a multiple of 3, then the nanotube is semiconducting with a very small band gap, otherwise the nanotube is a moderate semiconductor. Thus all armchair (n = m) nanotubes are metallic, and nanotubes (6,4), (9,1), etc. are semiconducting. Carbon nanotubes are not semi-metallic because the degenerate point (that point where the π [bonding] band meets the π* [anti-bonding] band, at which the energy goes to zero) is slightly shifted away from the K point in the Brillouin zone due to the curvature of the tube surface, casing hybridization between the σ* and π* anti-bonding bands, modifying the band dispersion.

The rule regarding metallic versus semiconductor behavior has exceptions, because curvature effects in small diameter tubes can strongly influence electrical properties. Thus, a (5,0) SWCNT that should be semiconducting in fact is metallic according to the calculations. Likewise, zigzag and chiral SWCNTs with small diameters that should be metallic have a finite gap (armchair nanotubes remain metallic). In theory, metallic nanotubes can carry an electric current density of 4×10^9 A/cm^2, which is more than 1,000 times greater than those of metals such as copper, where for copper interconnects current densities are limited by electromigration. Carbon nanotubes are thus being explored as interconnects, conductivity enhancing components in composite materials and many groups are attempting to commercialize highly conducting electrical wire assembled from individual carbon nanotubes. There are significant challenges to be overcome, however, such as undesired current saturation under voltage, the much more resistive nanotube-to-nanotube junctions and impurities, all of which lower the electrical conductivity of the macroscopic nanotube wires by orders of magnitude, as compared to the conductivity of the individual nanotubes.

Because of its nanoscale cross-section, electrons propagate only along the tube's axis. As a result, carbon nanotubes are frequently referred to as one-dimensional conductors. The maximum electrical conductance of a single-walled carbon nanotube is $2G_0$, where $G_0 = 2e^2/h$ is the conductance of a single ballistic quantum channel.

Due to the role of the π-electron system in determining the electronic properties of graphene, doping in carbon nanotubes differs from that of bulk crystalline semiconductors from the same group of the periodic table (e.g. silicon). Graphitic substitution of carbon atoms in the nanotube wall by boron or nitrogen dopants leads to p-type and n-type behavior, respectively, as would be expected in silicon. However, some non-substitutional (intercalated or adsorbed) dopants introduced into a carbon nanotube, such as alkali metals as well as electron-rich metallocenes, result in n-type conduction because they donate electrons to the π-electron system of the nanotube. By contrast, π-electron acceptors such as $FeCl_3$ or electron-deficient metallocenes function as p-type dopants since they draw π-electrons away from the top of the valence band.

Intrinsic superconductivity has been reported, although other experiments found no evidence of this, leaving the claim a subject of debate.

Optical

Carbon nanotubes have useful absorption, photoluminescence (fluorescence), and Raman spectroscopy properties. Spectroscopic methods offer the possibility of quick and non-destructive characterization of relatively large amounts of carbon nanotubes. There is a strong demand for such characterization from the industrial point of view: numerous parameters of the nanotube synthesis can be changed, intentionally or unintentionally, to alter the nanotube quality. As shown below, optical absorption, photoluminescence and Raman spectroscopies allow quick and reliable characterization of this "nanotube quality" in terms of non-tubular carbon content, structure (chirality) of the produced nanotubes, and structural defects. Those features determine nearly any other properties such as optical, mechanical, and electrical properties.

Carbon nanotubes are unique "one-dimensional systems" which can be envisioned as rolled single sheets of graphite (or more precisely graphene). This rolling can be done at different angles and curvatures resulting in different nanotube properties. The diameter typically varies in the range 0.4–40 nm (i.e. "only" ~100 times), but the length can vary ~10,000 times, reaching 55.5 cm. The nanotube aspect ratio, or the length-to-diameter ratio, can be as high as 132,000,000:1, which is unequalled by any other material. Consequently, all the properties of the carbon nanotubes relative to those of typical semiconductors are extremely anisotropic (directionally dependent) and tunable.

Whereas mechanical, electrical and electrochemical (supercapacitor) properties of the carbon nanotubes are well established and have immediate applications, the practical use of optical properties is yet unclear. The aforementioned tunability of properties is potentially useful in optics and photonics. In particular, light-emitting diodes (LEDs) and photo-detectors based on a single nanotube have been produced in the lab. Their unique feature is not the efficiency, which is yet relatively low, but the narrow selectivity in the wavelength of emission and detection of light and the possibility of its fine tuning through the nanotube structure. In addition, bolometer and optoelectronic memory devices have been realised on ensembles of single-walled carbon nanotubes.

Crystallographic defects also affect the tube's electrical properties. A common result is lowered conductivity through the defective region of the tube. A defect in armchair-type tubes (which can conduct electricity) can cause the surrounding region to become semiconducting, and single mon-atomic vacancies induce magnetic properties.

Thermal

All nanotubes are expected to be very good thermal conductors along the tube, exhibiting a property known as "ballistic conduction", but good insulators lateral to the tube axis. Measurements show that an individual SWNT has a room-temperature thermal conductivity along its axis of about 3500 $W·m^{-1}·K^{-1}$; compare this to copper, a metal well known for its good thermal conductivity, which transmits 385 $W·m^{-1}·K^{-1}$. An individual SWNT has a room-temperature thermal conductivity across its axis (in the radial direction) of about 1.52 $W·m^{-1}·K^{-1}$, which is about as thermally conductive as soil. Macroscopic assemblies of nanotubes such as films or fibres have reached up to 1500 $W·m^{-1}·K^{-1}$ so far. The temperature stability of carbon nanotubes is estimated to be up to 2800 °C in vacuum and about 750 °C in air.

Crystallographic defects strongly affect the tube's thermal properties. Such defects lead to phonon scattering, which in turn increases the relaxation rate of the phonons. This reduces the mean free path and reduces the thermal conductivity of nanotube structures. Phonon transport simulations indicate that substitutional defects such as nitrogen or boron will primarily lead to scattering of high-frequency optical phonons. However, larger-scale defects such as Stone Wales defects cause phonon scattering over a wide range of frequencies, leading to a greater reduction in thermal conductivity.

Synthesis

Techniques have been developed to produce nanotubes in sizable quantities, including arc discharge, laser ablation, high-pressure carbon monoxide disproportionation, and chemical vapor deposition (CVD). Most of these processes take place in a vacuum or with process gases. The CVD growth method is popular, as it yields high purity and has a high degree of control over diameter, length and morphology. Using particulate catalysts, large quantities of nanotubes can be synthesized by these methods; advances in catalysis and continuous growth are making CNTs more commercially viable.

Vertically aligned carbon nanotube arrays are also grown by thermal chemical vapor deposition. A substrate (quartz, silicon, stainless steel, etc.) is coated with a catalytic metal (Fe, Co, Ni) layer. Typically that layer is iron, and is deposited via sputtering to a thickness of 1–5 nm. A 10–50 nm underlayer of alumina is often also put down on the substrate first. This imparts controllable wetting and good interfacial properties. When the substrate is heated to the growth temperature (~700 °C), the continuous iron film breaks up into small islands each island then nucleates a carbon nanotube. The sputtered thickness controls the island size, and this in turn determines the nanotube diameter. Thinner iron layers drive down the diameter of the islands, and they drive down the diameter of the nanotubes grown. The amount of time that the metal island can sit at the growth temperature is limited, as they are mobile, and can merge into larger (but fewer) islands. Annealing at the growth temperature reduces the site density (number of CNT/mm^2) while increasing the catalyst diameter.

Chemical Modification

Carbon nanotubes can be functionalized to attain desired properties that can be used in a wide variety of applications. The two main methods of carbon nanotube functionalization are covalent and non-covalent modifications. Because of their hydrophobic nature, carbon nanotubes tend to agglomerate hindering their dispersion in solvents or viscous polymer melts. The resulting nanotube bundles or aggregates reduce the mechanical performance of the final composite. The surface of the carbon nanotubes can be modified to reduce the hydrophobicity and improve interfacial adhesion to a bulk polymer through chemical attachment.

Applications

Current

Current use and application of nanotubes has mostly been limited to the use of bulk nanotubes, which is a mass of rather unorganized fragments of nanotubes. Bulk nanotube materials may never achieve a tensile strength similar to that of individual tubes, but such composites may, nevertheless, yield strengths sufficient for many applications. Bulk carbon nanotubes have already been used as composite fibers in polymers to improve the mechanical, thermal and electrical properties of the bulk product.

- Easton-Bell Sports, Inc. have been in partnership with Zyvex Performance Materials, using CNT technology in a number of their bicycle components—including flat and riser handlebars, cranks, forks, seatposts, stems and aero bars.

- Zyvex Technologies has also built a 54' maritime vessel, the Piranha Unmanned Surface Vessel, as a technology demonstrator for what is possible using CNT technology. CNTs help improve the structural performance of the vessel, resulting in a lightweight 8,000 lb boat that can carry a payload of 15,000 lb over a range of 2,500 miles.

- Amroy Europe Oy manufactures Hybtonite carbon nanoepoxy resins where carbon nanotubes have been chemically activated to bond to epoxy, resulting in a composite material that is 20% to 30% stronger than other composite materials. It has been used for wind turbines, marine paints and a variety of sports gear such as skis, ice hockey sticks, baseball bats, hunting arrows, and surfboards.

Other current applications include:

- tips for atomic force microscope probes

- in tissue engineering, carbon nanotubes can act as scaffolding for bone growth

There is also ongoing research in using carbon nanotubes as a scaffold for diverse microfabrication techniques.

Potential

The strength and flexibility of carbon nanotubes makes them of potential use in controlling other nanoscale structures, which suggests they will have an important role in nanotechnology engineering. The highest tensile strength of an individual multi-walled carbon nanotube has been tested to be 63 GPa. Carbon nanotubes were found in Damascus steel from the 17th century, possibly helping to account for the legendary strength of the swords made of it. Recently, several studies have highlighted the prospect of using carbon nanotubes as building blocks to fabricate three-dimensional macroscopic (>1mm in all three dimensions) all-carbon devices. Lalwani et al. have reported a novel radical initiated thermal crosslinking method to fabricated macroscopic, free-standing, porous, all-carbon scaffolds using single- and multi-walled carbon nanotubes as building blocks. These scaffolds possess macro-, micro-, and nano- structured pores and the porosity can be tailored for specific applications. These 3D all-carbon scaffolds/architectures may be used for the fabrication of the next generation of energy storage, supercapacitors, field emission transistors, high-performance catalysis, photovoltaics, and biomedical devices and implants.

Large quantities of pure CNTs can be made into a fiber (a.k.a. filament) by wet spinning. The fiber is either directly spun from the synthesis pot or spun from pre-made dissolved CNTs. Individual fibers can be turned into a yarn. Apart from its strength and flexibility, the main advantage is making an electrically conducting yarn. The electronic properties of individual CNT fibers (i.e. bundle of individual CNT) is governed by the two-dimensional structure of CNTs.. At 300 K, CNT fibers have a resistivity one order of magnitude higher than the best electrical conductors.

CNT-based yarns are suitable for applications in energy and electrochemical water treatment when coated with an ion-exchange membrane. Also, CNT-based yarns could replace copper as a winding material. Pyrhönen et al. (2015) have built a motor using CNT winding.

Safety and Health

The National Institute for Occupational Safety and Health (NIOSH) is the leading United States federal agency conducting research and providing guidance on the occupational safety and health implications and applications of nanotechnology. Early scientific studies have indicated that some of these nanoscale particles may pose a greater health risk than the larger bulk form of these materials. In 2013, NIOSH published a Current Intelligence Bulletin detailing the potential hazards and recommended exposure limit for carbon nanotubes and fibers.

As of October 2016, single wall carbon nanotubes have been registered through the European Union's Registration, Evaluation, Authorization and Restriction of Chemicals (REACH) regulations, based on evaluation of the potentially hazardous properties of SWCNT. Based on this registration, SWCNT commercialization is allowed in the EU up to 10 metric tons. Currently, the type of SWCNT registered through REACH is limited to the specific type of single wall carbon nanotubes manufactured by OCSiAl, which submitted the application.

History

The true identity of the discoverers of carbon nanotubes is a subject of some controversy. A 2006 editorial written by Marc Monthioux and Vladimir Kuznetsov in the journal Carbon described the interesting and often-misstated origin of the carbon nanotube. A large percentage of academic and popular literature attributes the discovery of hollow, nanometer-size tubes composed of graphitic carbon to Sumio Iijima of NEC in 1991. He published a paper describing his discovery which initiated a flurry of excitement and could be credited by inspiring the many scientists now studying applications of carbon nanotubes. Though Iijima has been given much of the credit for discovering carbon nanotubes, it turns out that the timeline of carbon nanotubes goes back much further than 1991.

In 1952, L. V. Radushkevich and V. M. Lukyanovich published clear images of 50 nanometer diameter tubes made of carbon in the Soviet Journal of Physical Chemistry. This discovery was largely unnoticed, as the article was published in Russian, and Western scientists' access to Soviet press was limited during the Cold War. Monthioux and Kuznetsov mentioned in their Carbon editorial:

> The fact is, Radushkevich and Lukyanovich [..] should be credited for the discovery that carbon filaments could be hollow and have a nanometer- size diameter, that is to say for the discovery of carbon nanotubes.

In 1976, Morinobu Endo of CNRS observed hollow tubes of rolled up graphite sheets synthesised by a chemical vapour-growth technique. The first specimens observed would later come to be known as single-walled carbon nanotubes (SWNTs). Endo, in his early review of vapor-phase-grown carbon fibers (VPCF), also reminded us that he had observed a hollow tube, linearly extended with parallel carbon layer faces near the fiber core. This appears to be the observation of multi-walled carbon nanotubes at the center of the fiber. The mass-produced MWCNTs today are strongly related to the VPGCF developed by Endo. In fact, they call it the "Endo-process", out of respect for his early work and patents.

In 1979, John Abrahamson presented evidence of carbon nanotubes at the 14th Biennial Conference of Carbon at Pennsylvania State University. The conference paper described carbon nanotubes as carbon fibers that were produced on carbon anodes during arc discharge. A characterization of these fibers was given as well as hypotheses for their growth in a nitrogen atmosphere at low pressures.

In 1981, a group of Soviet scientists published the results of chemical and structural characterization of carbon nanoparticles produced by a thermocatalytical disproportionation of carbon monoxide. Using TEM images and XRD patterns, the authors suggested that their "carbon multi-layer tubular crystals" were formed by rolling graphene layers into cylinders. They speculated that by rolling graphene layers into a cylinder, many different arrangements of graphene hexagonal nets are possible. They suggested two possibilities of such arrangements: circular arrangement (armchair nanotube) and a spiral, helical arrangement (chiral tube).

In 1987, Howard G. Tennent of Hyperion Catalysis was issued a U.S. patent for the production of "cylindrical discrete carbon fibrils" with a "constant diameter between about 3.5 and about 70 nanometers…, length 10^2 times the diameter, and an outer region of multiple essentially continuous layers of ordered carbon atoms and a distinct inner core…."

Iijima's discovery of multi-walled carbon nanotubes in the insoluble material of arc-burned graphite rods in 1991 and Mintmire, Dunlap, and White's independent prediction that if single-walled carbon nanotubes could be made, then they would exhibit remarkable conducting properties helped create the initial buzz that is now associated with carbon nanotubes. Nanotube research accelerated greatly following the independent discoveries by Bethune at IBM and Iijima at NEC of single-walled carbon nanotubes and methods to specifically produce them by adding transition-metal catalysts to the carbon in an arc discharge. The arc discharge technique was well-known to produce the famed Buckminster fullerene on a preparative scale, and these results appeared to extend the run of accidental discoveries relating to fullerenes. The discovery of nanotubes remains a contentious issue. Many believe that Iijima's report in 1991 is of particular importance because it brought carbon nanotubes into the awareness of the scientific community as a whole.

Nanowire

A nanowire is a nanostructure, with the diameter of the order of a nanometer (10^{-9} meters). It can also be defined as the ratio of the length to width being greater than 1000. Alternatively, nanowires can be defined as structures that have a thickness or diameter constrained to tens of nanometers or less and an unconstrained length. At these scales, quantum mechanical effects are important — which coined the term "quantum wires". Many different types of nanowires exist, including superconducting (e.g. YBCO), metallic (e.g. Ni, Pt, Au), semiconducting (e.g. silicon nanowires (SiNWs), InP, GaN) and insulating (e.g. SiO_2, TiO_2). Molecular nanowires are composed of repeating molecular units either organic (e.g. DNA) or inorganic (e.g. $Mo_6S_{9-x}I_x$).

Overview

Crystalline 2×2-atom tin selenide nanowire grown inside a single-wall carbon nanotube (tube diameter ~1 nm).

A noise-filtered HRTEM image of a HgTe extreme nanowire embedded down the central pore of a SWCNT. The image is also accompanied by a simulation of the crystal structure.

Typical nanowires exhibit aspect ratios (length-to-width ratio) of 1000 or more. As such they are often referred to as one-dimensional (1-D) materials. Nanowires have many interesting properties that are not seen in bulk or 3-D (three-dimensional) materials. This is because electrons in nanowires are quantum confined laterally and thus occupy energy levels that are different from the traditional continuum of energy levels or bands found in bulk materials.

Peculiar features of this quantum confinement exhibited by certain nanowires manifest themselves in discrete values of the electrical conductance. Such discrete values arise from a quantum mechanical restraint on the number of electrons that can travel through the wire at the nanometer scale. These discrete values are often referred to as the quantum of conductance and are integer multiples of:

$$\frac{2e^2}{h} \simeq 77.41 \; \mu S$$

They are inverse of the well-known resistance unit h/e^2, which is roughly equal to 25812.8 ohms, and referred to as the von Klitzing constant R_K (after Klaus von Klitzing, the discoverer of exact quantization). Since 1990, a fixed conventional value R_{K-90} is accepted.

Examples of nanowires include inorganic molecular nanowires ($Mo_6S_{9-x}I_x$, $Li_2Mo_6Se_6$), which can have a diameter of 0.9 nm and be hundreds of micrometers long. Other important examples are based on semiconductors such as InP, Si, GaN, etc., dielectrics (e.g. SiO_2, TiO_2), or metals (e.g. Ni, Pt).

There are many applications where nanowires may become important in electronic, opto-electronic and nanoelectromechanical devices, as additives in advanced composites, for metallic interconnects in nanoscale quantum devices, as field-emitters and as leads for biomolecular nanosensors.

Synthesis of Nanowires

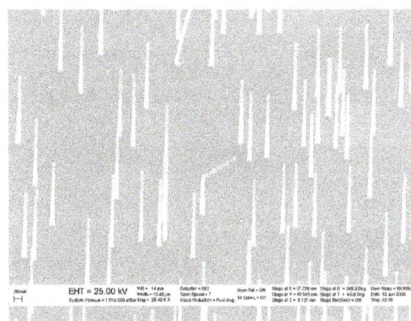

An SEM image of epitaxial nanowire heterostructures grown from catalytic gold nanoparticles.

There are two basic approaches to synthesizing nanowires: top-down and bottom-up. A top-down approach reduces a large piece of material to small pieces, by various means such as lithography or electrophoresis. A bottom-up approach synthesizes the nanowire by combining constituent adatoms. Most synthesis techniques use a bottom-up approach. Initial synthesis via either method may often be followed by a nanowire thermal treatment step, often involving a form of self-limiting oxidation, to fine tune the size and aspect ratio of the structures.

Nanowire production uses several common laboratory techniques, including suspension, electrochemical deposition, vapor deposition, and VLS growth. Ion track technology enables growing homogeneous and segmented nanowires down to 8 nm diameter.

Suspension

A suspended nanowire is a wire produced in a high-vacuum chamber held at the longitudinal extremities. Suspended nanowires can be produced by:

- The chemical etching of a larger wire

- The bombardment of a larger wire, typically with highly energetic ions

- Indenting the tip of a STM in the surface of a metal near its melting point, and then retracting it

VLS Growth

A common technique for creating a nanowire is vapor-liquid-solid method (VLS). This process can produce high-quality crystalline nanowires of many semiconductor materials, for example, VLS–grown single crystalline silicon nanowires (SiNWs) with smooth surfaces could have excellent properties, such as ultra-large elasticity. This method uses a source material from either laser ablated particles or a feed gas such as silane.

VLS synthesis requires a catalyst. For nanowires, the best catalysts are liquid metal (such as gold) nanoclusters, which can either be self-assembled from a thin film by dewetting, or purchased in colloidal form and deposited on a substrate.

The source enters these nanoclusters and begins to saturate them. On reaching supersaturation, the source solidifies and grows outward from the nanocluster. Simply turning off the source can adjust the final length of the nanowire. Switching sources while still in the growth phase can create compound nanowires with super-lattices of alternating materials.

A single-step vapour phase reaction at elevated temperature synthesises inorganic nanowires such as $Mo_6S_{9-x}I_x$. From another point of view, such nanowires are cluster polymers.

Solution-phase Synthesis

Solution-phase synthesis refers to techniques that grow nanowires in solution. They can produce nanowires of many types of materials. Solution-phase synthesis has the advantage that it can produce very large quantities, compared to other methods. In one technique, the polyol synthesis, ethylene glycol is both solvent and reducing agent. This technique is particularly versatile at producing nanowires of lead, platinum, and silver.

The supercritical fluid-liquid-solid growth method can be used to synthesize semiconductor nanowires, e.g., Si and Ge. By using metal nanocrystals as seeds, Si and Ge organometallic precursors are fed into a reactor filled with a supercritical organic solvent, such as toluene. Thermolysis results in degradation of the precursor, allowing release of Si or Ge, and dissolution into the metal nanocrystals. As more of the semiconductor solute is added from the supercritical phase (due to a concentration gradient), a solid crystallite precipitates, and a nanowire grows uniaxially from the nanocrystal seed.

Non-catalytic Growth

Nanowires can be also grown without the help of catalysts, which gives an advantage of pure nanowires and minimizes the number of technological steps. The simplest methods to obtain metal oxide nanowires use ordinary heating of the metals, e.g. metal wire heated with battery, by Joule heating in air can be easily done at home. The vast majority of nanowire-formation mechanisms are explained through the use of catalytic nanoparticles, which drive the nanowire growth and are either added intentionally or generated during the growth. However the mechanisms for catalyst-free growth of nanowires (or whiskers) were known from 1950s. Spontaneous nanowire formation by non-catalytic methods were explained by the dislocation present in specific directions or the growth anisotropy of various crystal faces. More recently, after microscopy advancement, the nanowire growth driven by screw dislocations or twin boundaries were demonstrated. The picture on the right shows a single atomic layer growth on the tip of CuO nanowire, observed by in situ TEM microscopy during the non-catalytic synthesis of nanowire.

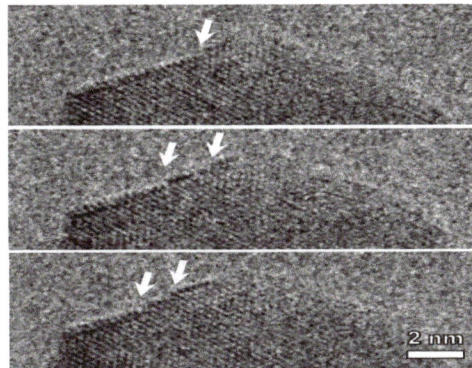

In situ observation of CuO nanowire growth.

Physics of Nanowires

Conductivity of Nanowires

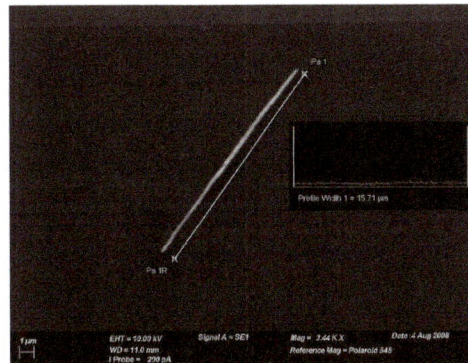

An SEM image of a 15 micrometer nickel wire.

Several physical reasons predict that the conductivity of a nanowire will be much less than that of the corresponding bulk material. First, there is scattering from the wire boundaries, whose effect will be very significant whenever the wire width is below the free electron mean free path of the bulk material. In copper, for example, the mean free path is 40 nm. Copper nanowires less than 40 nm wide will shorten the mean free path to the wire width.

Nanowires also show other peculiar electrical properties due to their size. Unlike single wall carbon nanotubes, whose motion of electrons can fall under the regime of ballistic transport (meaning the electrons can travel freely from one electrode to the other), nanowire conductivity is strongly influenced by edge effects. The edge effects come from atoms that lay at the nanowire surface and are not fully bonded to neighboring atoms like the atoms within the bulk of the nanowire. The unbonded atoms are often a source of defects within the nanowire, and may cause the nanowire to conduct electricity more poorly than the bulk material. As a nanowire shrinks in size, the surface atoms become more numerous compared to the atoms within the nanowire, and edge effects become more important.

Furthermore, the conductivity can undergo a quantization in energy: i.e. the energy of the electrons going through a nanowire can assume only discrete values, which are multiples of the conductance quantum $G = 2e^2/h$ (where e is the charge of the electron and h is the Planck constant).

The conductivity is hence described as the sum of the transport by separate channels of different quantized energy levels. The thinner the wire is, the smaller the number of channels available to the transport of electrons.

This quantization has been demonstrated by measuring the conductivity of a nanowire suspended between two electrodes while pulling it: as its diameter reduces, its conductivity decreases in a stepwise fashion and the plateaus correspond to multiples of G.

The quantization of conductivity is more pronounced in semiconductors like Si or GaAs than in metals, due to their lower electron density and lower effective mass. It can be observed in 25 nm wide silicon fins, and results in increased threshold voltage. In practical terms, this means that a MOSFET with such nanoscale silicon fins, when used in digital applications, will need a higher gate (control) voltage to switch the transistor on.

Welding Nanowires

To incorporate nanowire technology into industrial applications, researchers in 2008 developed a method of welding nanowires together: a sacrificial metal nanowire is placed adjacent to the ends of the pieces to be joined (using the manipulators of a scanning electron microscope); then an electric current is applied, which fuses the wire ends. The technique fuses wires as small as 10 nm.

For nanowires with diameters less than 10 nm, existing welding techniques, which require precise control of the heating mechanism and which may introduce the possibility of damage, will not be practical. Recently scientists discovered that single-crystalline ultrathin gold nanowires with diameters ~3–10 nm can be "cold-welded" together within seconds by mechanical contact alone, and under remarkably low applied pressures (unlike macro- and micro-scale cold welding process). High-resolution transmission electron microscopy and in situ measurements reveal that the welds are nearly perfect, with the same crystal orientation, strength and electrical conductivity as the rest of the nanowire. The high quality of the welds is attributed to the nanoscale sample dimensions, oriented-attachment mechanisms and mechanically assisted fast surface diffusion. Nanowire welds were also demonstrated between gold and silver, and silver nanowires (with diameters ~5–15 nm) at near room temperature, indicating that this technique may be generally applicable for ultrathin metallic nanowires. Combined with other nano- and microfabrication technologies, cold welding is anticipated to have potential applications in the future bottom-up assembly of metallic one-dimensional nanostructures.

Applications of Nanowires

Electronic Devices

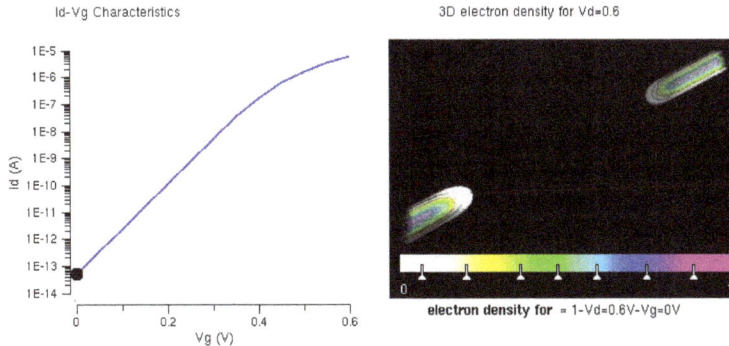

Atomistic simulation result for formation of inversion channel (electron density) and attainment of threshold voltage (IV) in a nanowire MOSFET. Note that the threshold voltage for this device lies around 0.45V.

Nanowires can be used for transistors. Transistors are used widely as fundamental building element in today's electronic circuits. As predicted by Moore's law, the dimension of transistors is shrinking smaller and smaller into nanoscale. One of the key challenges of building future nanoscale transistors is ensuring good gate control over the channel. Due to the high aspect ratio, if the gate dielectric is wrapped around the nanowire channel, we can get good control of channel electrostatic potential, thereby turning the transistor on and off efficiently.

To create active electronic elements, the first key step was to chemically dope a semiconductor nanowire. This has already been done to individual nanowires to create p-type and n-type semiconductors.

The next step was to find a way to create a p–n junction, one of the simplest electronic devices. This was achieved in two ways. The first way was to physically cross a p-type wire over an n-type wire. The second method involved chemically doping a single wire with different dopants along the length. This method created a p-n junction with only one wire.

After p-n junctions were built with nanowires, the next logical step was to build logic gates. By connecting several p-n junctions together, researchers have been able to create the basis of all logic circuits: the AND, OR, and NOT gates have all been built from semiconductor nanowire crossings.

In August 2012, researchers reported constructing the first NAND gate from undoped silicon nanowires. This avoids the problem of how to achieve precision doping of complementary nanocircuits, which is unsolved. They were able to control the Schottky barrier to achieve low-resistance contacts by placing a silicide layer in the metal-silicon interface.

It is possible that semiconductor nanowire crossings will be important to the future of digital computing. Though there are other uses for nanowires beyond these, the only ones that actually take advantage of physics in the nanometer regime are electronic.

In addition, nanowires are also being studied for use as photon ballistic waveguides as interconnects in quantum dot/quantum effect well photon logic arrays. Photons travel inside the tube, electrons travel on the outside shell.

When two nanowires acting as photon waveguides cross each other the juncture acts as a quantum dot.

Conducting nanowires offer the possibility of connecting molecular-scale entities in a molecular computer. Dispersions of conducting nanowires in different polymers are being investigated for use as transparent electrodes for flexible flat-screen displays.

Because of their high Young's moduli, their use in mechanically enhancing composites is being investigated. Because nanowires appear in bundles, they may be used as tribological additives to improve friction characteristics and reliability of electronic transducers and actuators.

Because of their high aspect ratio, nanowires are also uniquely suited to dielectrophoretic manipulation, which offers a low-cost, bottom-up approach to integrating suspended dielectric metal oxide nanowires in electronic devices such as UV, water vapor, and ethanol sensors.

Sensing of Proteins and Chemicals using Semiconductor Nanowires

In an analogous way to FET devices in which the modulation of conductance (flow of electrons/holes) in the semiconductor, between the input (source) and the output (drain) terminals, is controlled by electrostatic potential variation (gate-electrode) of the charge carriers in the device conduction channel, the methodology of a Bio/Chem-FET is based on the detection of the local change in charge density, or so-called "field effect", that characterizes the recognition event between a target molecule and the surface receptor.

This change in the surface potential influences the Chem-FET device exactly as a 'gate' voltage does, leading to a detectable and measurable change in the device conduction. When these devices are fabricated using semiconductor nanowires as the transistor element the binding of a chemical or biological species to the surface of the sensor can lead to the depletion or accumulation of charge carriers in the "bulk" of the nanometer diameter nanowire i.e. (small cross section available for conduction channels). Moreover, the wire, which serves as a tunable conducting channel, is in close contact with the sensing environment of the target, leading to a short response time, along with orders of magnitude increase in the sensitivity of the device as a result of the huge S/V ratio of the nanowires.

While several inorganic semiconducting materials such as Si, Ge, or metal oxides (e.g. In_2O_3, SnO_2, ZnO, etc.) have been used for the preparation of nanowires. Silicon nanowires are usually the material of choice when fabricating nanowire FET-based chemo/biosensors.

Several examples of the use of Silicon Nanowire(SiNW) sensing devices include the ultra sensitive, real-time sensing of biomarker proteins for cancer, detection of single virus particles, and the detection of nitro-aromatic explosive materials such as 2,4,6 Tri-nitrotoluene (TNT) in sensitives superior to these of canines. Silicon nanowires could also be used in their twisted form, as electromechanical devices, to measure intermolecular forces with great precision.

Limitations of Sensing with Silicon Nanowire FET Devices

Generally, the charges on dissolved molecules and macromolecules are screened by dissolved counterions, since in most cases molecules bound to the devices are separated from the sensor surface by approximately 2–12 nm (the size of the receptor proteins or DNA linkers bound to the sensor surface). As a result of the screening, the electrostatic potential that arises from charges on the analyte molecule decays exponentially toward zero with distance. Thus, for optimal sensing, the Debye length must be carefully selected for nanowire FET measurements. One approach of overcoming this limitation employs fragmentation of the antibody-capturing units and control over surface receptor density, allowing more intimate binding to the nanowire of the target protein. This approach proved useful for dramatically enhancing the sensitivity of cardiac biomarkers (e.g. Troponin) detection directly from serum for the diagnosis of acute myocardial infarction.

Nanofiber

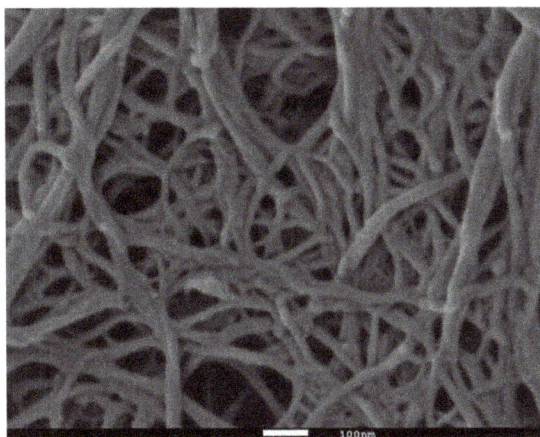

Example of a cellulose nanofiber network.

Nanofibers are fibers with diameters in the nanometer range. Nanofibers can be generated from different polymers and hence have different physical properties and application potentials. Examples of natural polymers include collagen, cellulose, silk fibroin, keratin, gelatin and polysaccharides such as chitosan and alginate. Examples of synthetic polymers include poly(lactic acid) (PLA), polycaprolactone (PCL), polyurethane (PU), poly(lactic-co-glycolic acid) (PLGA), poly(3-hydroxybutyrate-co-3-hydroxyvalerate) (PHBV), and poly(ethylene-co-vinylacetate) (PEVA). Polymer chains are connected via covalent bonds. The diameters of nanofibers depend on the type of polymer used and the method of production. All polymer nanofibers are unique for their large surface area-to-volume ratio, high porosity, appreciable mechanical strength, and flexibility in functionalization compared to their microfiber counterparts.

There exist many different methods to make nanofibers, including drawing, electrospinning, self-assembly, template synthesis, and thermal-induced phase separation. Electrospinning is the most commonly used method to generate nanofibers because of the straightforward setup, the ability to mass produce continuous nanofibers from various polymers, and the capability to generate ultrathin fibers with controllable diameters, compositions, and orientations. This flexibility allows for controlling the shape and arrangement of the fibers so that different structures (i.e. hollow, flat and ribbon shaped) can be fabricated depending on intended application purposes.

Nanofibers have many possible technological and commercial applications. They are used in tissue engineering, drug delivery, cancer diagnosis, lithium-air battery, optical sensors and air filtration.

History of Nanofiber Production

Nanofibers were first produced via electrospinning more than four centuries ago. Beginning with the development of the electrospinning method, English physicist William Gilbert (1544-1603) first documented the electrostatic attraction between liquids by preparing an experiment in which he observed a spherical water drop on a dry surface warp into a cone shape when it was held below an electrically charged amber. This deformation later came to be known as the Taylor cone. In 1882, English physicist Lord Rayleigh (1842-1919) analyzed the unstable states of liquid droplets that were electrically charged, and noted that the liquid was ejected in tiny jets when equilibrium was established between the surface tension and electrostatic force. In 1887, British physicist Charles Vernon Boys (1855-1944) published a manuscript about nanofiber development and production. In 1900, American inventor John Francis Cooley (1861-1903) filed the first modern electrospinning patent.

Anton Formhals was the first person to attempt nanofiber production between 1934 and 1944 and publish the first patent describing the experimental production of nanofibers. In 1966, Harold Simons published a patent for a device that could produce thin and light nanofiber fabrics with diverse motifs.

Only at the end of the 20th century have the words electrospinning and nanofiber become common language among scientists and researchers. Electrospinning continues to be developed today.

Synthesis Methods

Many chemical and mechanical techniques for preparing nanofibers exist.

Electrospinning

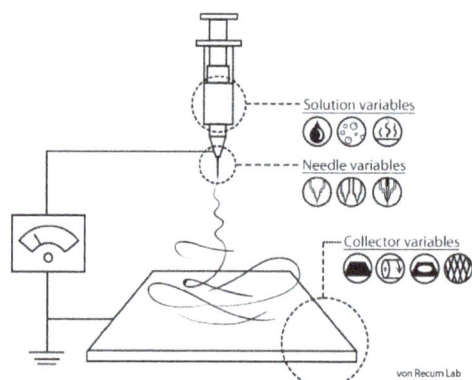

Diagram of a general set-up of electrospinning.

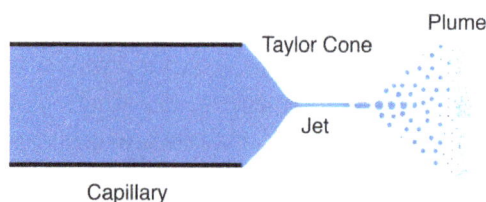

Taylor cone from which jet of polymer solution is ejected.

Electrospinning is the most commonly used method to fabricate nanofibers. The instruments necessary for electrospinning include a high voltage supplier, a capillary tube with a pipette or needle with a small diameter, and a metal collecting screen. One electrode is placed into the polymer solution and the other electrode is attached to the collector. An electric field is applied to the end of the capillary tube that contains the polymer solution held by its surface tension and forms a charge on the surface of the liquid. As the intensity of the electric field increases, the hemispherical surface of the fluid at the tip of the capillary tube elongates to form a conical shape known as the Taylor cone. A critical value is attained upon further increase in the electric field in which the repulsive electrostatic force overcomes the surface tension and the charged jet of fluid is ejected from the tip of the Taylor cone. The discharged polymer solution jet is unstable and elongates as a result, allowing the jet to become very long and thin. Charged polymer fibers solidifies with solvent evaporation. Randomly-oriented nanofibers are collected on the collector. Nanofibers can also be collected in a highly-aligned fashion by using specialized collectors such as the rotating drum, metal frame, or a two-parallel plates system. Parameters such as jet stream movement and polymer concentration have to be controlled to produce nanofibers with uniform diameters and morphologies.

The electrospinning technique transforms many types of polymers into nanofibers. An electrospun nanofiber network resembles the extracellular matrix (ECM) well. This resemblance is a major advantage of electrospinning because it opens up the possibility of mimicking the ECM with regards to fiber diameters, high porosity, and mechanical properties. Electrospinning is being further developed for mass production of one-by-one continuous nanofibers.

Thermal-induced Phase Separation

Thermal-induced phase separation separates a homogenous polymer solution into a multi-phase system via thermodynamic changes. The procedure involves five steps: polymer dissolution, liquid-liquid or liquid-solid phase separation, polymer gelation, extraction of solvent from the gel with water, and freezing and freeze-drying under vacuum. Thermal-induced phase separation method is widely used to generate scaffolds for tissue regeneration.

The homogenous polymer solution in the first step is thermodynamically unstable and tends to separate into polymer-rich and polymer-lean phases under appropriate temperature. Eventually after solvent removal, the polymer-rich phase solidifies to form the matrix and the polymer-lean phase develops into pores. Next, two types of phase separation can be carried out on the polymer solution depending on the desired pattern. Liquid-liquid separation is usually used to form bicontinuous phase structures while solid-liquid phase separation is used to form crystal structures. The gelation step plays a crucial role in controlling the porous morphology of the nanofibrous matrices. Gelation is influenced by temperature, polymer concentration, and solvent properties. Temperature regulates the structure of the fiber network: low gelation temperature results in formation of nanoscale fiber networks while high gelation temperature leads to the formation of a platelet-like structure. Polymer concentration affects fiber properties: an increase in polymer concentration decreases porosity and increases mechanical properties such as tensile strength. Solvent properties influence morphology of the scaffolds. After gelation, gel is placed in distilled water for solvent exchange. Afterwards, the gel is removed from the water and goes through freezing and freeze-drying. It is then stored in a desiccator until characterization.

Drawing

The drawing method makes long single strands of nanofibers one at a time. The pulling process is accompanied by solidification that converts the dissolved spinning material into a solid fiber. A cooling step is necessary in the case of melt spinning and evaporation of solvent in the case of dry spinning. A limitation, however, is that only a viscoelastic material that can undergo extensive deformations while possessing sufficient cohesion to survive the stresses developed during pulling can be made into nanofibers through this process.

Template Synthesis

The template synthesis method uses a nanoporous membrane template composed of cylindrical pores of uniform diameter to make fibrils (solid nanofiber) and tubules (hollow nanofiber). This method can be used to prepare fibrils and tubules of many types of materials, including metals, semiconductors and electronically conductive polymers. The uniform pores allow for control of the dimensions of the fibers so nanofibers with very small diameters can be produced through this method. However, a drawback of this method is that it cannot make continuous nanofibers one at a time.

Self-assembly

The self-assembly technique is used to generate peptide nanofibers and peptide amphiphiles. The method was inspired by the natural folding process of amino acid residues to form proteins with unique three-dimensional structures. The self-assembly process of peptide nanofibers involves various driving forces such as hydrophobic interactions, electrostatic forces, hydrogen bonding and van der Waals forces and is influenced by external conditions such as ionic strength and pH.

Polymer Materials

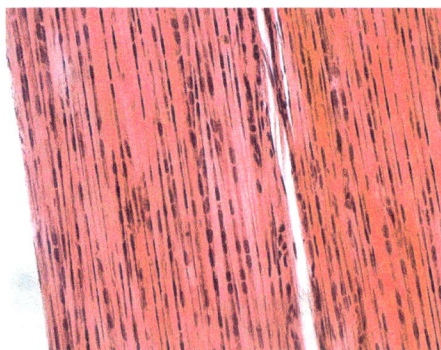

Collagen fibers in a cross-sectional area of dense connective tissue.

Due to their high porosity and large surface area-to-volume ratio, nanofibers are widely used to construct scaffolds for biological applications. Major examples of natural polymers used in scaffold production are collagen, cellulose, silk fibroin, keratin, gelatin and polysaccharides such as chitosan and alginate. Collagen is a natural extracellular component of many connective tissues. Its fibrillary structure, which varies in diameter from 50-500 nm, is important for cell recognition, attachment, proliferation and differentiation. Using type I collagen nanofibers produced via electrospinning, Shih et al. found that the engineered collagen scaffold showed an increase in cell adhesion and decrease in cell migration with increasing fiber diameter. Using silk scaffolds as a guide for growth for bone tissue regeneration, Kim et al. observed complete bone union after 8 weeks and complete healing of defects after 12 weeks whereas the control in which the bone did not have the scaffold displayed limited mending of defects in the same time period. Similarly, keratin, gelatin, chitosan and alginate demonstrate excellent biocompatibility and bioactivity in scaffolds.

However, cellular recognition of natural polymers can easily initiate an immune response. Consequently, synthetic polymers such as poly(lactic acid) (PLA), polycaprolactone (PCL), polyurethane (PU), poly(lactic-co-glycolic acid) (PLGA), poly(L-lactide) (PLLA), and poly(ethylene-co-vinylacetate) (PEVA) have been developed as alternatives for integration into scaffolds. Being biodegradable and biocompatible, these synthetic polymers can be used to form matrices with a fiber diameter within the nanometer range. Out of these synthetic polymers, PCL has generated considerable enthusiasm among researchers. PCL is a type of biodegradable polyester that can be prepared via ring-opening polymerization of ε-caprolactone using catalysts. It shows low toxicity, low cost and slow degradation. PCL can be combined with other materials such as gelatin, collagen, chitosan, and calcium phosphate to improve the differentiation and proliferation capacity (2, 17). PLLA is another popular synthetic polymer. PLLA is well known for its superior mechanical properties, biodegradability and biocompatibility. It shows efficient cell migration ability due to its high spatial interconnectivity, high porosity and controlled alignment. A blend of PLLA and PLGA scaffold matrix has shown proper biomimetic structure, good mechanical strength and favorable bioactivity.

Applications

Tissue Engineering

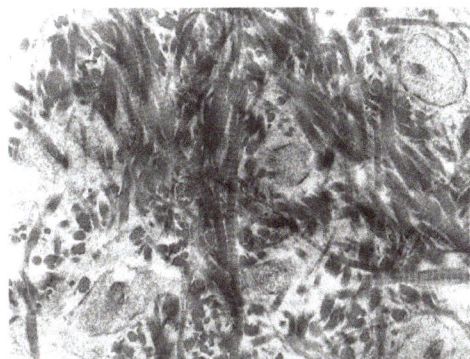

Bone matrix composed of collagen fibrils. Nanofiber scaffolds are able to mimic such structure.

In tissue engineering, a highly porous artificial extracellular matrix is needed to support and guide cell growth and tissue regeneration. Natural and synthetic biodegradable polymers have been used to create such scaffolds.

Nanofiber scaffolds are used in bone tissue engineering to mimic the natural extracellular matrix of the bones. The bone tissue is arranged either in a compact or trabecular pattern and composed of organized structures that vary in length from the centimeter range all the way to the nanometer scale. Nonmineralized organic component (i.e. type 1 collagen), mineralized inorganic component (i.e. hydroxyapatite), and many other noncollagenous matrix proteins (i.e. glycoproteins and proteoglycans) make up the nanocomposite structure of the bone ECM. The organic collagen fibers and the inorganic mineral salts provide flexibility and toughness, respectively, to ECM.

Although the bone is a dynamic tissue that can self-heal upon minor injuries, it cannot regenerate after experiencing large defects such as bone tumor resections and severe nonunion fractures because it lacks the appropriate template. Currently, the standard treatment is autografting which involves obtaining the donor bone from a non-significant and easily accessible site (i.e. iliac crest) in the patient own body and transplanting it into the defective site. Transplantation of autologous bone has the best clinical outcome because it integrates reliably with the host bone and can avoid complications with the immune system. But its use is limited by its short supply and donor site morbidity associated with the harvest procedure. Furthermore, autografted bones are avascular and hence are dependent on diffusion for nutrients, which affects their viability in the host. The grafts can also be resorbed before osteogenesis is complete due to high remodeling rates in the body. Another strategy for treating severe bone damage is allografting which transplants bones harvested from a human cadaver. However, allografts introduce the risk of disease and infection in the host.

Bone tissue engineering presents a versatile response to treat bone injuries and deformations. Nanofibers produced via electrospinning mimics the architecture and characteristics of natural extracellular matrix particularly well. These scaffolds can be used to deliver bioactive agents that promote tissue regeneration. These bioactive materials should ideally be osteoinductive, osteoconductive, and osseointegratable. Bone substitute materials intended to replace autologous or allogeneic bone consist of bioactive ceramics, bioactive glasses, and biological and synthetic polymers. The basis of bone tissue engineering is that the materials will be resorbed and replaced over time by the body's own newly regenerated biological tissue.

Tissue engineering is not only limited to the bone: a large amount of research is devoted to cartilage, ligament, skeletal muscle, skin, blood vessel, and neural tissue engineering as well.

Drug Delivery

Successful delivery of therapeutics to the intended target largely depends on the choice of the drug carrier. The criteria for an ideal drug carrier include maximum effect upon delivery of the drug to the target organ, evasion of the immune system of the body in the process of reaching the organ, retention of the therapeutic molecules from preparatory stages to the final delivery of the drug, and proper release of the drug for exertion of the intended therapeutic effect.

Nanofibers have grabbed the attention of researchers as a favorable carrier candidate. Natural polymers such as gelatin and alginate make for good fabrication biomaterials for carrier nanofibers because of their biocompatibility and biodegradability that result in no harm to the tissue of the host and no toxic accumulation in the human body, respectively. Additionally, the physical properties of the nanofibers are well suited for drug delivery system. Due to their cylindrical morphology, they possess a high surface area-to-volume ratio. As a result, the fibers possess high drug-loading capacity and can release therapeutic molecules over a large surface area in the medium. Whereas surface area to volume ratio can only be controlled by adjusting the radius for spherical vesicles, nanofibers have more degrees of freedom in controlling the ratio by varying both the length and the cross-sectional radius. This adjustability is important for their application in drug delivery system in which the functional parameters need to be precisely controlled.

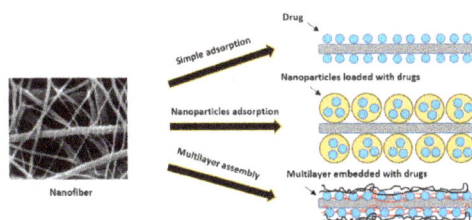

Drugs and biopolymers can be loaded onto nanofibers via simple adsorption, nanoparticles adsorption, and multilayer assembly.

Antibiotics and anticancer drugs have been successfully encapsulated in electrospun nanofibers by adding the drug into the polymer solution prior to electrospinning. Ignatova et al. showed that various antibiotics such as tetracycline hydrochloride, ciprofloxacin and levofloxacin can be incorporated into the synthetic polymer solution that can be electrospun to produce fibers that can be delivered to the site of interest. Drugs and other biopolymers such as DNA can be loaded onto the surfaces of nanofibers via simple adsorption, nanoparticles adsorption, and multilayer assembly. Surface-loaded nanofiber scaffolds are useful as adhesion barriers between internal organs and tissues post-surgery. Adhesion occurs during the healing process and can bring on complications such as chronic pain and reoperation failure. Nanofibers can separate the surgery-operated site from nearby tissues and organs while simultaneously deliver therapeutics such as antibiotics that are physically adsorbed onto their surfaces. Drugs can also be constructed in the form of nanoparticles and attached to the surface of nanofibers, allowing for loading of high concentration of the drugs in a given nanofiber. Rujitanaroj et al showed that nanofibers functionalized with silver nanoparticles can be used as effective antibiotics against bacteria such as E. coli, Pseudomonas aeruginosa, Staphylococcus aureus, and methicillin-resistant Staphylococcus aureus found in burn wounds. Finally, charged biopolymers such as DNA can be incorporated into the multilayer assembly of nanofibers. Such modification can provide diverse drug releasing surface profiles of nanofibers for different medical contexts.

Cancer Diagnosis

Although pathologic examination is the current standard method for molecular characterization in testing for the presence of biomarkers in tumors, these single-sample analyses fail to account for the diverse genomic nature of tumors. Considering the invasive nature, psychological stress, and the financial burden resulting from repeated tumor biopsies in patients, biomarkers that could be judged through minimally invasive procedures, such as blood draws, constitute an opportunity for progression in precision medicine.

Liquid biopsy is an option that is becoming increasingly popular as an alternative to solid tumor biopsy. This is simply a blood draw that contains circulating tumor cells (CTCs) which are shed into the bloodstream from solid tumors. Patients with metastatic cancer are more likely to have detectable CTCs in the bloodstream but CTCs also exist in patients with localized diseases. It has been found that the number of CTCs present in the bloodstream of patients with metastatic prostate and colorectal cancer is prognostic of the overall survival of tumors. CTCs also have been demonstrated to inform prognosis in earlier stages of the disease.

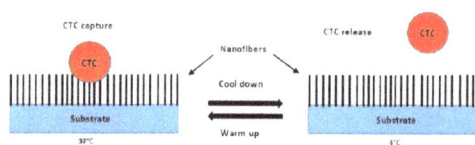

CTC capture and release mechanism of third generation Thermoresponsive Chip.

Recently, Ke et al. developed a NanoVelcro chip that captures the CTCs from the blood samples. When blood is passed through the chip, the nanofibers coated with protein antibodies bind to the proteins expressed on the surface of cancer cells and act like Velcro to trap CTCs for analysis. The NanoVelcro CTC assays underwent three generations of development. The first generation NanoVelcro Chip was created for CTC enumeration for cancer prognosis, staging, and dynamic monitoring. The second generation NanoVelcro-LCM was developed for single-cell CTC isolation. The individually isolated CTCs can be subjected to single-CTC genotyping. The third generation Thermoresponsive Chip allowed for CTC purification. The nanofiber polymer brushes undergo temperature-dependent conformational changes to capture and release CTCs.

Lithium-air Battery

Among many advanced electrochemical energy storage devices, rechargeable lithium-air batteries are of particular interest due to their considerable energy storing capacities and high power densities. As the battery is being used, lithium ions combine with oxygen from the air to form particles of lithium oxides, which attach to carbon fibers on the electrode. During recharging, the lithium oxides separate again into lithium and oxygen which is released back into the atmosphere. This conversion sequence is highly inefficient because there is significant voltage difference of more than 1.2 volts between the output voltage and the charging voltage of the battery meaning that approximately 30% of the electrical energy is lost as heat when the battery is charging. Also the large volume changes resulting from continuous conversion of oxygen between its gaseous and solid state puts stress on the electrode and limits its lifetime.

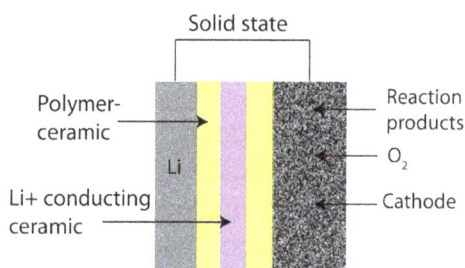

Schematic of a lithium-air battery. For the nanofiber-based lithium-air battery, the cathode would be made up of carbon nanofibers.

The performance of these batteries depends on the characteristics of the material that makes up the cathode. Carbon materials have been widely used as cathodes because of their excellent electrical conductivities, large surface areas, and chemical stability. Especially relevant for lithium-air batteries, carbon materials act as substrates for supporting metal oxides. Binder-free electrospun carbon nanofibers are particularly good potential candidates to be used in electrodes in lithium-oxygen batteries because they have no binders, have open macroporous structures, have carbons that support and catalyze the oxygen reduction reactions, and have versatility.

Zhu et al. developed a novel cathode that can store lithium and oxygen in the electrode they named nanolithia which is a matrix of carbon nanofibers periodically embedded with cobalt oxide. These cobalt oxides provide stability to the normally unstable superoxide-containing nanolithia. In this design, oxygen is stored as LiO_2 and does not convert between gaseous and solid forms during charging and discharging. When the battery is discharging, lithium ions in nanolithia and react with superoxide oxygen the matrix to form Li_2O_2, and Li_2O. The oxygen remains in its solid state as it transitions among these forms. The chemical reactions of these transitions provide electrical energy. During charging, the transitions occur in reverse.

Optical Sensors

Polymer optical fibers have generated increasing interest in recent years. Because of low cost, ease of handling, long wavelength transparency, great flexibility, and biocompatibility, polymer optical fibers show great potential for short-distance networking, optical sensing and power delivery.

Electrospun nanofibers are particularly well-suitable for optical sensors because sensor sensitivity increases with increasing surface area per unit mass. Optical sensing works by detecting ions and molecules of interest via fluorescence quenching mechanism. Wang et al. successfully developed nanofibrous thin film optical sensors for metal ion (Fe^{3+} and Hg^{2+}) and 2,4-dinitrotoluene (DNT) detection using the electrospinning technique.

Quantum dots show useful optical and electrical properties, including high optical gain and photochemical stability. A variety of quantum dots have been successfully incorporated into polymer nanofibers. Meng et al. showed that quantum dot-doped polymer nanofiber sensor for humidity detection shows fast response, high sensitivity, and long-term stability while requiring low power consumption.

Kelly et al. developed a sensor that warns first responders when the carbon filters in their respirators have become saturated with toxic fume particles. The respirators typically contain activated charcoal that traps airborne toxins. As the filters become saturated, chemicals begin to pass through and render the respirators useless. In order to easily determine when the filter is spent, Kelly and his team developed a mask equipped with a sensor composed of carbon nanofibers assembled into repeating structures called photonic crystals that reflect specific wavelengths of light. The sensors exhibit an iridescent color that changes when the fibers absorb toxins.

Air Filtration

Paints and protective coatings on furniture contain volatile organic compounds such as toluene and formaldehyde.

Electrospun nanofibers are useful for removing volatile organic compounds (VOC) from the atmopshere. Scholten et al. showed that adsorption and desorption of VOC by electrospun nanofibrous membrane were faster than the rates of conventional activated carbon.

Airborne contamination in the personnel cabins of mining equipment is of concern to the mining workers, mining companies, and government agencies such as the Mine Safety and Health Administration (MSHA). Recent work with mining equipment manufacturers and the MSHA has shown that nanofiber filter media can reduce cabin dust concentration to a greater extent compared to standard cellulose filter media.

Nanoparticle

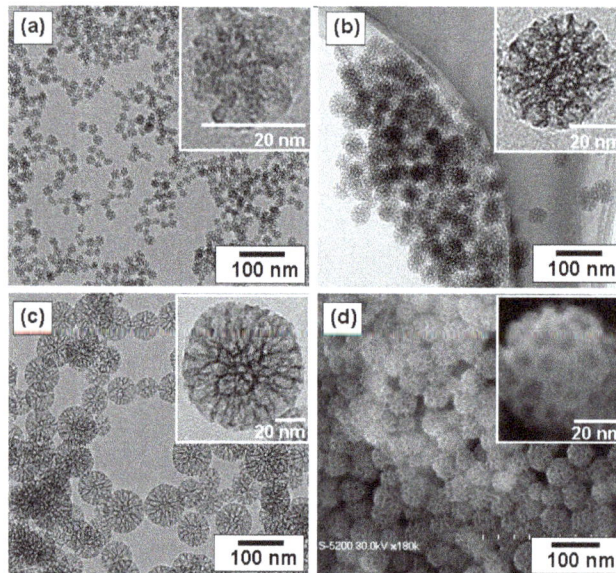

TEM (a, b, and c) images of prepared mesoporous silica nanoparticles with mean outer diameter: (a) 20nm, (b) 45nm, and (c) 80nm. SEM (d) image corresponding to (b). The insets are a high magnification of mesoporous silica particle.

Nanoparticles are particles between 1 and 100 nanometers in size. In nanotechnology, a particle is defined as a small object that behaves as a whole unit with respect to its transport and properties. Particles are further classified according to diameter. Ultrafine particles are the same as nanoparticles and between 1 and 100 nanometers in size, fine particles are sized between 100 and 2,500 nanometers, and coarse particles cover a range between 2,500 and 10,000 nanometers. Scientific research on nanoparticles is intense as they have many potential applications in medicine, physics, optics, and electronics. The U.S. National Nanotechnology Initiative offers government funding focused on nanoparticle research.

Definition

The term "nanoparticle" is not usually applied to individual molecules; it usually refers to inorganic materials.

The reason for the synonymous definition of nanoparticles and ultrafine particles is that, during the 1970s and 80s, when the first thorough fundamental studies with "nanoparticles" were underway in the USA (by Granqvist and Buhrman) and Japan, (within an ERATO Project) they were called "ultrafine particles" (UFP). However, during the 1990s before the National Nanotechnology Initiative was launched in the USA, the new name, "nanoparticle," had become more common (for example, see the same senior author's paper 20 years later addressing the same issue, lognormal distribution of sizes). Nanoparticles can exhibit size-related properties significantly different from those of either fine particles or bulk materials.

Nanoclusters have at least one dimension between 1 and 10 nanometers and a narrow size distribution. Nanopowders are agglomerates of ultrafine particles, nanoparticles, or nanoclusters. Nanometer-sized single crystals, or single-domain ultrafine particles, are often referred to as nanocrystals.

Background

Although nanoparticles are associated with modern science, they have a long history. Nanoparticles were used by artisans as far back as Rome in the fourth century in the famous Lycurgus cup made of dichroic glass as well as the ninth century in Mesopotamia for creating a glittering effect on the surface of pots.

In modern times, pottery from the Middle Ages and Renaissance often retains a distinct gold- or copper-colored metallic glitter. This luster is caused by a metallic film that was applied to the transparent surface of a glazing. The luster can still be visible if the film has resisted atmospheric oxidation and other weathering.

The luster originates within the film itself, which contains silver and copper nanoparticles dispersed homogeneously in the glassy matrix of the ceramic glaze. These nanoparticles are created by the artisans by adding copper and silver salts and oxides together with vinegar, ochre, and clay on the surface of previously-glazed pottery. The object is then placed into a kiln and heated to about 600 °C in a reducing atmosphere.

In the heat the glaze softens, causing the copper and silver ions to migrate into the outer layers of the glaze. There the reducing atmosphere reduced the ions back to metals, which then came together forming the nanoparticles that give the color and optical effects.

Luster technique showed that ancient craftsmen had a sophisticated empirical knowledge of materials. The technique originated in the Muslim world. As Muslims were not allowed to use gold in artistic representations, they sought a way to create a similar effect without using real gold. The solution they found was using luster.

Michael Faraday provided the first description, in scientific terms, of the optical properties of nanometer-scale metals in his classic 1857 paper. In a subsequent paper, the author (Turner) points out that: "It is well known that when thin leaves of gold or silver are mounted upon glass and heated to a temperature that is well below a red heat (~500 °C), a remarkable change of properties takes place, whereby the continuity of the metallic film is destroyed. The result is that white light is now freely transmitted, reflection is correspondingly diminished, while the electrical resistivity is enormously increased."

Uniformity

The chemical processing and synthesis of high-performance technological components for the private, industrial, and military sectors requires the use of high-purity ceramics (oxide ceramics, such as aluminium oxide or copper(II) oxide), polymers, glass-ceramics, and composite materials, as metal carbides (SiC), nitrides (Aluminum nitrides, Silicon nitride), metals (Al, Cu), non-metals (graphite, carbon nanotubes) and layered (Al + Aluminium carbonate, Cu + C). In condensed bodies formed from fine powders, the irregular particle sizes and shapes in a typical powder often lead to non-uniform packing morphologies that result in packing density variations in the powder compact.

Uncontrolled agglomeration of powders due to attractive van der Waals forces can also give rise to microstructural heterogeneity. Differential stresses that develop as a result of non-uniform drying shrinkage are directly related to the rate at which the solvent can be removed, and thus highly dependent upon the distribution of porosity. Such stresses have been associated with a plastic-to-brittle transition in consolidated bodies, and can yield to crack propagation in the unfired body if not relieved.

In addition, any fluctuations in packing density in the compact as it is prepared for the kiln are often amplified during the sintering process, yielding inhomogeneous densification. Some pores and other structural defects associated with density variations have been shown to play a detrimental role in the sintering process by growing and thus limiting end-point densities. Differential stresses arising from inhomogeneous densification have also been shown to result in the propagation of internal cracks, thus becoming the strength-controlling flaws.

Inert gas evaporation and inert gas deposition are free many of these defects due to the distillation (cf. purification) nature of the process and having enough time to form single crystal particles, however even their non-aggreated deposits have lognormal size distribution, which is typical with nanoparticles. The reason why modern gas evaporation techniques can produce a relatively narrow size distribution is that aggregation can be avoided. However, even in this case, random residence times in the growth zone, due to the combination of drift and diffusion, result in a size distribution appearing lognormal.

It would, therefore, appear desirable to process a material in such a way that it is physically uniform with regard to the distribution of components and porosity, rather than using particle size distributions that will maximize the green density. The containment of a uniformly dispersed assembly of strongly interacting particles in suspension requires total control over interparticle forces. Monodisperse nanoparticles and colloids provide this potential.

Monodisperse powders of colloidal silica, for example, may therefore be stabilized sufficiently to ensure a high degree of order in the colloidal crystal or polycrystalline colloidal solid that results from aggregation. The degree of order appears to be limited by the time and space allowed for longer-range correlations to be established. Such defective polycrystalline colloidal structures would appear to be the basic elements of submicrometer colloidal materials science and, therefore, provide the first step in developing a more rigorous understanding of the mechanisms involved in microstructural evolution in high performance materials and components.

Properties

Nanoparticles are of great scientific interest as they are, in effect, a bridge between bulk materials and atomic or molecular structures. A bulk material should have constant physical properties regardless of its size, but at the nano-scale size-dependent properties are often observed. Thus, the properties of materials change as their size approaches the nanoscale and as the percentage of the surface in relation to the percentage of the volume of a material becomes significant. For bulk materials larger than one micrometer (or micron), the percentage of the surface is insignificant in relation to the volume in the bulk of the material. *The interesting and sometimes unexpected properties of nanoparticles are therefore largely due to the large surface area of the material, which dominates the contributions made by the small bulk of the material.*

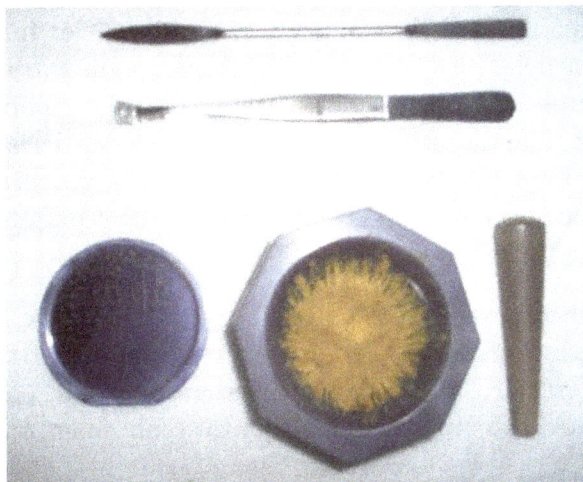

Silicon nanopowder

1 kg of particles of 1 mm³ has the same surface area as 1 mg of particles of 1 nm³

Nanoparticles often possess unexpected optical properties as they are small enough to confine their electrons and produce quantum effects. For example, gold nanoparticles appear deep-red to black in solution. Nanoparticles of yellow gold and grey silicon are red in color. Gold nanoparticles melt at much lower temperatures (~300 °C for 2.5 nm size) than the gold slabs (1064 °C);. Absorption of solar radiation is much higher in materials composed of nanoparticles than it is in thin films of continuous sheets of material. In both solar PV and solar thermal applications, controlling the size, shape, and material of the particles, it is possible to control solar absorption.

Other size-dependent property changes include quantum confinement in semiconductor particles, surface plasmon resonance in some metal particles and superparamagnetism in magnetic materials. What would appear ironic is that the changes in physical properties are not always desirable. Ferromagnetic materials smaller than 10 nm can switch their magnetisation direction using room temperature thermal energy, thus making them unsuitable for memory storage.

Suspensions of nanoparticles are possible since the interaction of the particle surface with the solvent is strong enough to overcome density differences, which otherwise usually result in a material either sinking or floating in a liquid.

The high surface area to volume ratio of nanoparticles provides a tremendous driving force for diffusion, especially at elevated temperatures. Sintering can take place at lower temperatures, over shorter time scales than for larger particles. In theory, this does not affect the density of the final product, though flow difficulties and the tendency of nanoparticles to agglomerate complicates matters. Moreover, nanoparticles have been found to impart some extra properties to various day to day products. For example, the presence of titanium dioxide nanoparticles imparts what we call the self-cleaning effect, and, the size being nano-range, the particles cannot be observed. Zinc oxide particles have been found to have superior UV blocking properties compared to its bulk substitute. This is one of the reasons why it is often used in the preparation of sunscreen lotions, is completely photostable and toxic.

Clay nanoparticles when incorporated into polymer matrices increase reinforcement, leading to stronger plastics, verifiable by a higher glass transition temperature and other mechanical property tests. These nanoparticles are hard, and impart their properties to the polymer (plastic). Nanoparticles have also been attached to textile fibers in order to create smart and functional clothing.

Metal, dielectric, and semiconductor nanoparticles have been formed, as well as hybrid structures (e.g., core–shell nanoparticles). Nanoparticles made of semiconducting material may also be labeled quantum dots if they are small enough (typically sub 10 nm) that quantization of electronic energy levels occurs. Such nanoscale particles are used in biomedical applications as drug carriers or imaging agents.

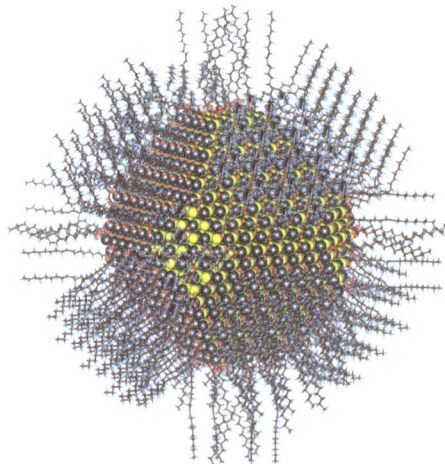

Semiconductor nanoparticle (quantum dot) of lead sulfide with complete passivation by oleic acid, oleyl amine and hydroxyl ligands (size ~5nm)

Semi-solid and soft nanoparticles have been manufactured. A prototype nanoparticle of semi-solid nature is the liposome. Various types of liposome nanoparticles are currently used clinically as delivery systems for anticancer drugs and vaccines.

Nanoparticles with one half hydrophilic and the other half hydrophobic are termed Janus particles and are particularly effective for stabilizing emulsions. They can self-assemble at water/oil interfaces and act as solid surfactants.

Hydrogel nanoparticles made of N-isopropylacrylamide hydrogel core shell can be dyed with affinity baits, internally. These affinity baits allow the nanoparticles to isolate and remove undesirable proteins while enhancing the target analytes.

Synthesis

There are several methods for creating nanoparticles, including gas condensation, attrition, chemical precipitation, pyrolysis and hydrothermal synthesis. In attrition, macro- or micro-scale particles are ground in a ball mill, a planetary ball mill, or other size-reducing mechanism. The resulting particles are air classified to recover nanoparticles. In pyrolysis, a vaporous precursor (liquid or gas) is forced through an orifice at high pressure and burned. The resulting solid (a version of soot) is air classified to recover oxide particles from by-product gases. Traditional pyrolysis often results in aggregates and agglomerates rather than single primary particles. Ultrasonic nozzle spray pyrolysis (USP) on the other hand aids in preventing agglomerates from forming.

A thermal plasma can deliver the energy to vaporize small micrometer-size particles. The thermal plasma temperatures are in the order of 10,000 K, so that solid powder easily evaporates. Nanoparticles are formed upon cooling while exiting the plasma region. The main types of the thermal plasma torches used to produce nanoparticles are dc plasma jet, dc arc plasma, and radio frequency (RF) induction plasmas. In the arc plasma reactors, the energy necessary for evaporation and reaction is provided by an electric arc formed between the anode and the cathode. For example, silica sand can be vaporized with an arc plasma at atmospheric pressure, or thin aluminum wires can be vaporized by exploding wire method. The resulting mixture of plasma gas and silica vapour can be rapidly cooled by quenching with oxygen, thus ensuring the quality of the fumed silica produced.

In RF induction plasma torches, energy coupling to the plasma is accomplished through the electromagnetic field generated by the induction coil. The plasma gas does not come in contact with electrodes, thus eliminating possible sources of contamination and allowing the operation of such plasma torches with a wide range of gases including inert, reducing, oxidizing, and other corrosive atmospheres. The working frequency is typically between 200 kHz and 40 MHz. Laboratory units run at power levels in the order of 30–50 kW, whereas the large-scale industrial units have been tested at power levels up to 1 MW. As the residence time of the injected feed droplets in the plasma is very short, it is important that the droplet sizes are small enough in order to obtain complete evaporation. The RF plasma method has been used to synthesize different nanoparticle materials, for example synthesis of various ceramic nanoparticles such as oxides, carbours/carbides, and nitrides of Ti and Si.

Inert-gas condensation is frequently used to make nanoparticles from metals with low melting points. The metal is vaporized in a vacuum chamber and then supercooled with an inert gas stream. The supercooled metal vapor condenses into nanometer-size particles, which can be entrained in the inert gas stream and deposited on a substrate or studied in situ.

Nanoparticles can also be formed using radiation chemistry. Radiolysis from gamma rays can create strongly active free radicals in solution. This relatively simple technique uses a minimum number of chemicals. These including water, a soluble metallic salt, a radical scavenger (often a secondary alcohol), and a surfactant (organic capping agent). High gamma doses on the order of 10^4 Gray are required. In this process, reducing radicals will drop metallic ions down to the zero-valence state. A scavenger chemical will preferentially interact with oxidizing radicals to prevent the re-oxidation of the metal. Once in the zero-valence state, metal atoms begin to coalesce into particles. A chemical surfactant surrounds the particle during formation and regulates its growth. In sufficient concentrations, the surfactant molecules stay attached to the particle. This prevents it from dissociating or forming clusters with other particles. Formation of nanoparticles using the radiolysis method allows for tailoring of particle size and shape by adjusting precursor concentrations and gamma dose.

Sol-gel

The sol-gel process is a wet-chemical technique (also known as chemical solution deposition) widely used recently in the fields of materials science and ceramic engineering. Such methods are used primarily for the fabrication of materials (typically a metal oxide) starting from a chemical solution (*sol*, short for solution), which acts as the precursor for an integrated network (or *gel*) of either discrete particles or network polymers.

Typical precursors are metal alkoxides and metal chlorides, which undergo hydrolysis and poly-condensation reactions to form either a network "elastic solid" or a colloidal suspension (or dispersion) – a system composed of discrete (often amorphous) submicrometer particles dispersed to various degrees in a host fluid. Formation of a metal oxide involves connecting the metal centers with oxo (M-O-M) or hydroxo (M-OH-M) bridges, therefore generating metal-oxo or metal-hydroxo polymers in solution. Thus, the sol evolves toward the formation of a gel-like diphasic system containing both a liquid phase and solid phase whose morphologies range from discrete particles to continuous polymer networks.

In the case of the colloid, the volume fraction of particles (or particle density) may be so low that a significant amount of fluid may need to be removed initially for the gel-like properties to be recognized. This can be accomplished in a number of ways. The most simple method is to allow time for sedimentation to occur, and then pour off the remaining liquid. Centrifugation can also be used to accelerate the process of phase separation.

Removal of the remaining liquid (solvent) phase requires a drying process, which typically causes shrinkage and densification. The rate at which the solvent can be removed is ultimately determined by the distribution of porosity in the gel. The ultimate microstructure of the final component will clearly be strongly influenced by changes implemented during this phase of processing. Afterward, a thermal treatment, or firing process, is often necessary in order to favor further poly-condensation and enhance mechanical properties and structural stability via final sintering, densification, and grain growth. One of the distinct advantages of using this methodology as opposed to the more traditional processing techniques is that densification is often achieved at a much lower temperature.

The precursor sol can be either deposited on a substrate to form a film (e.g., by dip-coating or spin-coating), cast into a suitable container with the desired shape (e.g., to obtain a monolithic ceramics, glasses, fibers, membranes, aerogels), or used to synthesize powders (e.g., microspheres, nanospheres). The sol-gel approach is a cheap and low-temperature technique that allows for the fine control of the product's chemical composition. Even small quantities of dopants, such as organic dyes and rare earth metals, can be introduced in the sol and end up uniformly dispersed in the final product. It can be used in ceramics processing and manufacturing as an investment casting material, or as a means of producing very thin films of metal oxides for various purposes. Sol-gel derived materials have diverse applications in optics, electronics, energy, space, (bio)sensors, medicine (e.g., controlled drug release) and separation (e.g., chromatography) technology.

Colloids

The term colloid is used primarily to describe a range of mixtures which have solid particles dispersed in a liquid medium. The term applies only if the particles are larger than atomic dimensions but small enough to exhibit Brownian motion. If the particles are large enough, their dynamic behavior in any given period of time in suspension would be governed by forces of gravity and sedimentation. If the particles are small enough, their irregular motion in suspension can be attributed to the collective bombardment of a myriad of thermally agitated molecules in the liquid suspending medium, as described originally by Albert Einstein in his dissertation. Einstein showed evidence that water was made up of discrete molecules by characterizing the observed erratic particle behavior using the theory of Brownian motion, with sedimentation being a possible result. The critical size range (or particle diameter) typically ranges from nanometers (10^{-9} m) to micrometers (10^{-6} m).

Morphology

Nanostars of vanadium(IV) oxide

Scientists have taken to naming their particles after the real-world shapes that they might represent. Nanospheres, nanochains, nanoreefs, nanoboxes and more have appeared in the literature. These morphologies sometimes arise spontaneously as an effect of a templating or directing agent present in the synthesis such as miscellar emulsions or anodized alumina pores, or from the innate crystallographic growth patterns of the materials themselves. Some of these morphologies may serve a purpose, such as long carbon nanotubes used to bridge an electrical junction, or just a scientific curiosity like the stars shown at right.

Amorphous particles usually adopt a spherical shape (due to their microstructural isotropy), whereas the shape of anisotropic microcrystalline whiskers corresponds to their particular crystal habit. At the small end of the size range, nanoparticles are often referred to as clusters. Spheres, rods, fibers, and cups are just a few of the shapes that have been grown. The study of fine particles is called micromeritics.

Characterization

The majority of nanoparticle characterization techniques are light-based, but a non-optical nanoparticle characterization technique called Tunable Resistive Pulse Sensing (TRPS) has been developed that enables the simultaneous measurement of size, concentration and surface charge for a wide variety of nanoparticles. This technique, which applies the Coulter Principle, allows for particle-by-particle quantification of these three nanoparticle characteristics with high resolution.

Functionalization

The surface coating of nanoparticles determines many of their properties, notably stability, solubility, and targeting. A coating that is multivalent or polymeric confers high stability. Functionalized nanomaterial-based catalysts can be used for catalysis of many known organic reactions.

Surface Coating for Biological Applications

For biological applications, the surface coating should be polar to give high aqueous solubility and prevent nanoparticle aggregation. In serum or on the cell surface, highly charged coatings promote non-specific binding, whereas polyethylene glycol linked to terminal hydroxyl or methoxy groups repel non-specific interactions. Nanoparticles can be linked to biological molecules that can act as address tags, to direct the nanoparticles to specific sites within the body, specific organelles within the cell, or to follow specifically the movement of individual protein or RNA molecules in living cells. Common address tags are monoclonal antibodies, aptamers, streptavidin or peptides. These targeting agents should ideally be covalently linked to the nanoparticle and should be present in a controlled number per nanoparticle. Multivalent nanoparticles, bearing multiple targeting groups, can cluster receptors, which can activate cellular signaling pathways, and give stronger anchoring. Monovalent nanoparticles, bearing a single binding site, avoid clustering and so are preferable for tracking the behavior of individual proteins.

Red blood cell coatings can help nanoparticles evade the immune system.

Safety

Nanoparticles present possible dangers, both medically and environmentally. Most of these are due to the high surface to volume ratio, which can make the particles very reactive or catalytic. They are also able to pass through cell membranes in organisms, and their interactions with biological systems are relatively unknown. However, it is unlikely the particles would enter the cell nucleus, Golgi complex, endoplasmic reticulum or other internal cellular components due to the particle size and intercellular agglomeration. A recent study looking at the effects of ZnO nanoparticles on human immune cells has found varying levels of susceptibility to cytotoxicity. There are concerns that pharmaceutical companies, seeking regulatory approval for nano-reformulations of existing medicines, are relying on safety data produced during clinical studies of the earlier, pre-reformulation version of the medicine. This could result in regulatory bodies, such as the FDA, missing new side effects that are specific to the nano-reformulation.

Whether cosmetics and sunscreens containing nanomaterials pose health risks remains largely unknown at this stage. However considerable research has demonstrated that zinc nanoparticles are not absorbed into the bloodstream in vivo.

Concern has also been raised over the health effects of respirable nanoparticles from certain combustion processes. As of 2013 the U.S. Environmental Protection Agency was investigating the safety of the following nanoparticles:

- Carbon Nanotubes: Carbon materials have a wide range of uses, ranging from composites for use in vehicles and sports equipment to integrated circuits for electronic components. The interactions between nanomaterials such as carbon nanotubes and natural organic matter strongly influence both their aggregation and deposition, which strongly affects their transport, transformation, and exposure in aquatic environments. In past research, carbon nanotubes exhibited some toxicological impacts that will be evaluated in various environmental settings in current EPA chemical safety research. EPA research will provide data, models, test methods, and best practices to discover the acute health effects of carbon nanotubes and identify methods to predict them.

- Cerium oxide: Nanoscale cerium oxide is used in electronics, biomedical supplies, energy, and fuel additives. Many applications of engineered cerium oxide nanoparticles naturally disperse themselves into the environment, which increases the risk of exposure. There is ongoing exposure to new diesel emissions using fuel additives containing CeO_2 nanoparticles, and the environmental and public health impacts of this new technology are unknown. EPA's chemical safety research is assessing the environmental, ecological, and health implications of nanotechnology-enabled diesel fuel additives.

- Titanium dioxide: Nano titanium dioxide is currently used in many products. Depending on the type of particle, it may be found in sunscreens, cosmetics, and paints and coatings. It is also being investigated for use in removing contaminants from drinking water.

- Nano Silver: Nano silver is being incorporated into textiles, clothing, food packaging, and other materials to eliminate bacteria. EPA and the U.S. Consumer Product Safety Commission are studying certain products to see whether they transfer nano-size silver particles in real-world scenarios. EPA is researching this topic to better understand how much nano-silver children come in contact with in their environments.

- Iron: While nano-scale iron is being investigated for many uses, including "smart fluids" for uses such as optics polishing and as a better-absorbed iron nutrient supplement, one of its more prominent current uses is to remove contamination from groundwater. This use, supported by EPA research, is being piloted at a number of sites across the country.

Laser Applications

The use of nanoparticles in laser dye-doped poly(methyl methacrylate) (PMMA) laser gain media was demonstrated in 2003 and it has been shown to improve conversion efficiencies and to decrease laser beam divergence. Researchers attribute the reduction in beam divergence to improved dn/dT characteristics of the organic-inorganic dye-doped nanocomposite. The optimum composition reported by these researchers is 30% w/w of SiO_2 (~ 12 nm) in dye-doped PMMA.

Medicinal Applications

- Liposome

- Dendrimer

- Iron oxide nanoparticles

- Nanomedicine

- Polymer-drug conjugate

- Polymeric nanoparticle

References

- Lambin, P. (1996). "Atomic structure and electronic properties of bent carbon nanotubes". Synth. Met. 77: 249–1254. doi:10.1016/0379-6779(96)80097

- MacNaught, Alan D.; Wilkinson, Andrew R., eds. (1997). Compendium of Chemical Terminology: IUPAC Recommendations (2nd ed.). Blackwell Science. ISBN 0865426848

- Ma, K.L. (2011). "Electronic transport properties of junctions between carbon nanotubes and graphene nanoribbons". European Physical Journal B. 83: 487–492. doi:10.1140/epjb/e2011-20313-9

- Bowman-James, Kristin (2005). "Alfred Werner Revisited: The Coordination Chemistry of Anions". Accounts of Chemical Research. 38 (8): 671–678. PMID 16104690. doi:10.1021/ar040071

- Vogel, A.I., Tatchell, A.R., Furnis, B.S., Hannaford, A.J. and P.W.G. Smith. Vogel's Textbook of Practical Organic Chemistry, 5th Edition. Prentice Hall, 1996. ISBN 0-582-46236-3

- "CDC – NIOSH Publications and Products – Current Intelligence Bulletin 65: Occupational Exposure to Carbon Nanotubes and Nanofibers (2013–145)". www.cdc.gov. Retrieved 2017-02-01

- Gavrilova, A. L.; Bosnich, B., "Principles of Mononucleating and Binucleating Ligand Design", Chem. Rev. 2004, volume 104, 349-383. doi:10.1021/cr020604g

- Takeuchi, K.; Hayashi, T.; Kim, Y. A.; Fujisawa, K. and Endo, M. (February 2014) "The state-of-the-art science and applications of carbon nanotubes", nanojournal.ifmo.ru

- Mintmire, J.W.; Dunlap, B.I.; White, C.T. (1992). "Are Fullerene Tubules Metallic?". Phys. Rev. Lett. 68 (5): 631–634. Bibcode:1992PhRvL..68..631M. PMID 10045950. doi:10.1103/PhysRevLett.68.631. Cotton, Frank Albert; Geoffrey Wilkinson; Carlos A. Murillo (1999). Advanced Inorganic Chemistry. Wiley-Interscience. p. 1355. ISBN 978-0471199571

- Agam, M. A.; Guo, Q (2007). "Electron Beam Modification of Polymer Nanospheres". Journal of Nanoscience and Nanotechnology. 7 (10): 3615–9. doi:10.1166/jnn.2007.814. PMID 18330181

- Dekker, C. (1999). "Carbon nanotubes as molecular quantum wires". Physics Today. 52 (5): 22–28. Bibcode:1999PhT....52e..22D. doi:10.1063/1.882658

- Miessler, Gary L.; Paul J. Fischer; Donald Arthur Tarr (2013). Inorganic Chemistry. Prentice Hall. p. 696. ISBN 978-0321811059

- "Toxic Nanoparticles Might be Entering Human Food Supply, MU Study Finds". University of Missouri. 22 August 2013. Retrieved 23 August 2013

- Vines T, Faunce T (2009). "Assessing the safety and cost-effectiveness of early nanodrugs". Journal of law and medicine. 16 (5): 822–45. PMID 19554862

- Hartwig, J. F. Organotransition Metal Chemistry, from Bonding to Catalysis; University Science Books: New York, 2010. ISBN 1-891389-53

- Davidson, Keay (17 May 2006). "FDA urged to limit nanoparticle use in cosmetics and sunscreens". San Francisco Chronicle. Retrieved 20 April 2007

- Chernozatonskii, L.A. (1992). "Carbon nanotube connectors and planar jungle gyms". Physics Letters A. 172: 173–176. doi:10.1016/0375-9601(92)90978-u

- Howard, V. (2009). "Statement of Evidence: Particulate Emissions and Health (An Bord Plenala, on Proposed Ringaskiddy Waste-to-Energy Facility)." Retrieved 26 April 2011.

- Menon, M. (1997). "Carbon Nanotube "T Junctions": Nanoscale Metal-Semiconductor-Metal Contact Devices". Physical Review Letters. 79: 4453– 4456. doi:10.1103/physrevlett.79.4453

Green Synthesis in Nanotechnology

As opposed to physical and chemical methods, green synthesis is non-toxic, cost-effective, and eco-friendly. However, it has its drawbacks. It is time-consuming and proves difficult to control size, shape and crystallinity. Nonetheless, it can optimize metal ion concentration, pH, temperature etc., and hence is widely used in biotechnology. This section discusses an important method of nanobiotechnology in a critical manner providing key analysis to the subject matter.

Green Synthesis

Nanotechnology has attracted a great deal of attention over the recent years due to applications in various fields such as energy, medicine, electronic and space industries. There are two strategies for nanoparticle synthesis: top-down and bottom-up. In top-down approach, bulk material is broken down into small pieces gradually and in bottom-up approach, atoms and molecules are brought together to synthesize nano-sized particles. Bottom-up approach is generally used for biological synthesis of nanoparticles.

Due to their small size and high surface area nanoparticle have characteristic physical, chemical, mechanical, electronic, electrical, optical, magnetic, thermal, dielectric and biological properties not possessed by their larger sized counterparts. Optoelectronic, physiochemical and all other properties of nanoparticles are determined by the shape, size and monodispersity of the particle. These characteristics depend upon the method of synthesis of nanoparticles. Physical and Chemical methods used now for production of nanoparticles though lead to monodisperse nanoparticles, but they are less stable and various toxic chemicals are used. The use of toxic chemicals and non-polar solvents in synthesis leads to the inability to use nanoparticles in clinical fields. Therefore, development of clean, non-toxic, biocompatible and eco-friendly method for synthesis of nanoparticles deserves recognition. Even though biological synthesis of nanoparticles is considered cost effective, safe, environment-friendly and sustainable, it has various drawbacks. The culturing of microorganisms is time-consuming and it is difficult to have fine control over shape, size and crystallinity. The particles are not monodisperse and the rate of production is slow. These are the various problems which have vexed the biological synthesis of nanoparticles. But optimization of factors involved like pH, temperature, metal ion concentration, and the strain of the microbe used has given hope for large scale application of biological synthesis. Moreover genetically engineered strains which express the reducing agent maximally can be used in the future which will provide better control over the shape and size of nanoparticles. Interaction between microbes and metals has been known for long and is used in bioremediation, biomineralization, bioleaching and biocorrosion but it its use in the synthesis of nanoparticles is a recent discovery and lot of study is required before it can be put to practical use.

Nanoparticle Synthesis by Bacteria

Microorganisms often produce inorganic materials of nano-size either extarcellularly or intracellularly. Microbial systems are able to detoxify heavy metals by virtue of their ability to reduce the metal ions or precipitate the soluble toxic ions into insoluble non-toxic metal nanoparticles. A great deal of study has been carried out on synthesis of nanoparticles by prokaryotic bacteria since they are the easiest organisms to handle and can be manipulated most easily. Bacteria are able to form nanoparticles both intracellularly via bioaccumulation and extarcellularly on the cell wall using its enzymes. Intracellular nanoparticles are of a fixed size with less monodispersity than extracellular particles. Hence, extracellular production has more commercial applications in various fields. Since monodispersity is the major factor in usefulness of nanoparticles, biological processes must be designed in such a way to ensure maximum monodispersity.

To obtain intracellular particles from bacteria requires further processing steps like ultrasound treatment or reaction with suitable detergents. This property can be exploited for extraction of precious metals from mine wastes and the metal nanoparticles can also be used as catalysts. When cell wall reductive enzymes or secreted enzymes are involved in the reduction of metal ions then it is logical to find the metal nanoparticles outside the cell. The extracellular nanoparticles have wider applications in the field of optoelectronics, bioimaging and sensor technology than intracellular particles.

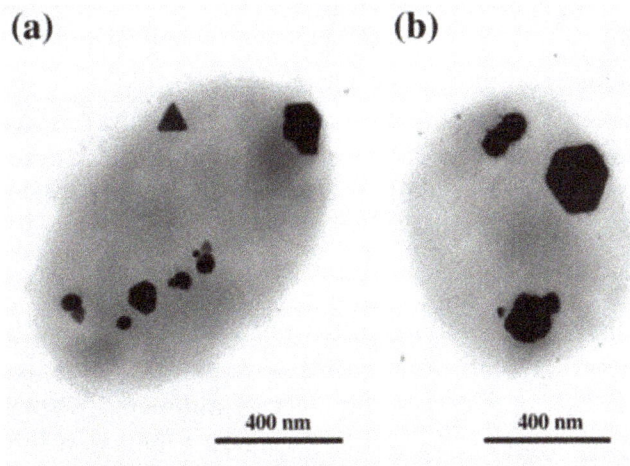

In one of the earliest study in this field, it was found that a silver-resistant bacterial strain is lated from silver mines, Pseudomonas Stutzeri AG259 was able to accumulate silver nanoparticles in its periplasmic space with the size ranging between 36-45 nanometers along with some silver sulfide. It was also noted that when this bacteria is placed in a concentrated aqueous solution of silver nitrate larger nanoparticles of up to 200 nanometers with defined morphology were formed. Cell growth and metal incubation conditions may be the reason for disparity in the size. The exact mechanism of formation of nanoparticles by this species of bacteria is yet to be understood. The ability of microorganisms to resist high concentration of toxic metal ions may result from specific reaction mechanisms. These may include efflux systems, extracellular precipitation by secreted/cell wall enzymes, alteration of solubility and toxicity by changing the oxidation state of the metal or a absence of certain transport systems. Nanocrystalline silver can be recovered from the bacteria by thermally treating the bacteria to yield a carbonaceous nanomaterial. This material is composed of five percent by weight silver nanoparticles and rest dry biomass and has found applications in thin-film coating materials. Bacillus subtilis is able to reduce gold to form octahedral gold nanoparticles of size between 5 to 25 nanometers when incubated along with gold chloride solution. These are examples of generation of nanoparticles by bacteria in their natural settings. Some bacteria are able to produce nanoparticles in presence of concentrated metal solution and appropriate incubation settings even when naturally they do not face such conditions and do not produce any nanoparticles naturally. The exposure of some Lactobacillus strains (found in buttermilk), like Lactobacillus sp.A09, to silver and gold solution lead to the formation of nanoparticles. It can also be used for the production of gold-silver alloy. Similarly, dried cell mass of Corynebactetium sp.SH09 was able to produce silver nanoparticles with silver diamine complex at 60 degree centigrade after 72 hours. It is believed that organic matrix of the cell provides peptides that contain amino acid moieties that serve as a nucleation point for the start of nanoparticle formation. Silver precipitating peptides were found to have capable of reducing aqueous silver face centered cubic structured silver crystals. The exact mechanism of the formation is not yet known and further research needs to be carried out.

A prokaryotic bacterium Rhodopseudomonas capsulata, was found to deposit gold nanoparticles of 10-20 nanometers at 7 pH and room temperature extarcellularly. As the pH of the solution was changed various nanoparticles with different sizes and geometries (like triangular and spherical at 4.0 pH) were formed. It was found experimentally that cell free extract of Rhodopseudomonas Capsulata can also be used for production of gold nanoparticles. SDS-PAGE analysis of the extract demonstrated the involvement of one or more proteins (14-98 kDa) in the reduction of gold and capping of gold nanoparticles. Similarly, silver nanoparticles can be produced extarcellularly using Enterobacter culture supernatant. These bacteria secrete enzymes in their culture solutions which are able to reduce silver and assist in the formation silver nanoparticles. UV-visible spectroscopy and Transmission Electron Microscopy of the solution estimates the size of particles between 28-122 nanometers with the average size of 52.5 nanometers.

In addition to gold and silver much attention has been focused on synthesis protocols of semiconductors (quantum dots) like cadmium sulfide, zinc sulfide and lead sulfide. These luminescent quantum dots are emerging as a new set of materials with important applications in cell imaging and biosensing, based on the conjugation between biorecognition molecules and quantum dots. On conjugation these can be visualized easily because of their luminescence. Clostridium thermoaceticum was found to deposit CdS nanoparticles on the cell surface as well as in the solution in presence of cadmium chloride and cysteine hydrochloride. Possibly, cysteine hydrochloride acts as the source of sulfur. When Klebsiella aerogenes is exposed cadmium ions in the growth medium it forms cadmium sulfide nanoparticles of 20-200 nanometers deposited on the cell surface. Escherichia coli when incubated with cadmium chloride and sodium sulfide forms intracellular cadmium sulfide nanoparticles in wurtzite crystal phase. Experiments show that the growth phase of the cells affect the formation rate of nanoparticles and is 20 times more I stationary phase than in late logarithmic phase. Zinc sulfide nanoparticles are formed by the family Desulfobacteriaceae in their natural settings by a complex mechanism. This can be used to bring down the zinc concentration in drinking water to below acceptable levels. Magnetic iron sulphide nanoparticles can also be produced by sulfite reducing bacteria.

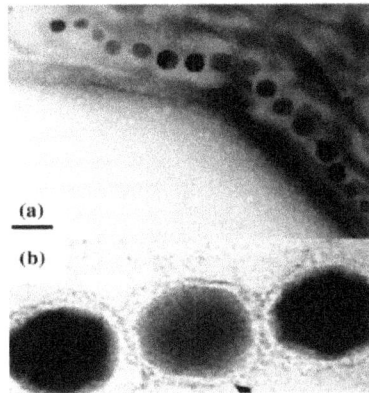

(a)

(b)

Studies have shown that microserophilic bacteria Aquaspirilium magnetotacticum was able to form single domain magnetite nanoparticles with octahedral geometry. Marine magnetotactic bacterium MV-1 isolated from sulfide rich sediments was also able to form magnetite nanoparticles (parallelepiped) of dimensions 40*40*60 nanometers. A thermophilic fermentative bacterial strain TOR-39 was also able to form single domain (<12 nanometers) magnetite octahedral nanoparticles exclusively outside the cell. Another bacteria Magnetospirillium magnetotacticum was also able to form single domain nanoparticles which are subsequently assembled into folded chain and flux closure assemblies. Their 2-D arrangement is responsible for the head-tail assembly. Magnetization studies have shown that the magnetite nanoparticles are not superparamagnetic. All magnetotactic bacteria producing magnetic particles intracellularly contain another organelle called magnetosomes which are comprised iron mineral crystals protected by a membrane vesicle. This membrane is most likely the structure that holds the crystal at a particular location in the cell as well as serves as the starting point (nucleation point) for nanoparticle formation. Possibly, biological systems exert control over growth of the crystals using the same magnetosmal membrane. The bacteria producing the magnetic nanoparticles can be separated from the culture using micro electromagnets. After they are separated from the culture they can be subjected to lysis to leave the crystals at desired locations. Magnetite nanoparticles have also been produced by non magnetotactic bacteria Geobacter metallireducens GS-15 isolated from the sediments of Potomac River. The amorphous ferric oxide acts as a terminal electron acceptor during organic matter oxidation and magnetite crystals are formed.

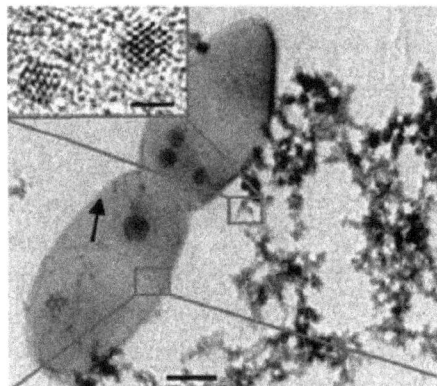

Stenotrophomonas maltophilia SELTE02, a strain isolated from soil near selenium accumulator legume was found capable of transforming selenite to elemental selenium and accumulate it inside the cells as well as deposit it extarcellularly. A facultative anaerobe Enterobacter cloacea SLD1a-1

and Desulfovibrio desulfricans have also been found capable to reduce selenite to selenium. Desulfovibrio desulfricans NCIMB 8307 was able to generate palladium and platinum nanoparticles.

Nanoparticle Synthesis of Fungi

Fungi are a relatively recent addition to the list of microorganisms used for nanoparticle synthesis. Fungi are more beneficial than other microorganisms in many ways. They grow in the form of mycelial mesh which helps them to bear flow pressure and agitation and other conditions to which microbes are subjected to in a bioreactor used for large scale production. Though they require better precision and care to grow but they are easier to handle and manipulate. They secrete enzymes in large amounts. The nanoparticles are generally generated extarcellularly (sometimes intracellular production does occur) they are devoid of various impurities from the cell and can be used directly.

The use of eukaryotic organisms for nanoparticle synthesis was first demonstrated by use of Verticillium sp. for the synthesis of gold nanoparticles. In this experiment, gold nanoparticles were reported on the surface and cytoplasmic membrane of the fungal mycelia. Due to the formation of gold nanoparticles the mycelial mass attains a typical purple color demonstrating intracellular generation. TEM analysis shows that particles of well-defined geometry like triangular, hexagonal or spherical shape were formed on the cell wall and quasi-hexagonal morphology were formed of the cytoplasmic membrane. The fungal biomass on exposure to silver nitrate solution was also found generate silver nanoparticles intracellularly. The powder diffraction indicates the crystal nature of both the nanoparticles. The exact mechanism for the synthesis of nanoparticles by Verticillium is not yet known. It is thought that the first step is the interaction between positively charged metal ions and negatively charged carboxylates on the enzymes present in fungal cell wall and adhesion of the metal ions to the surface as a result of this interaction. The enzymes reduce these metal ions to elemental metal which serve as nucleation sites and further growth is carried out by subsequent reduction and accumulation. The ability of Verticillium to grow and replicate even after exposure to metal ions demonstrate their ability to be used commercially for production of nanoparticles.

A plant pathogenic fungus, Fusarium oxysporum has also been studied extensively. It was found that it was able to generate gold and silver nanoparticles extarcellularly rapidly. This was observed from the fact that the supernatant changed its color but the mycelial mass retained its original color. Moreover the fungal extract was also able to generate gold and silver nanoparticles. It is believed that the fungus releases reductases in the solution which are responsible for the reduction of metal ions. This makes in-vitro generation of nanoparticles using an enzyme/cell extract based process possible. It was recently discovered that when F. oxysporum is exposed to equimolar solutions of hydrogen tetrachloroaurate (III) and silver nitrate led to the production of gold-silver alloy. The presence of only one plasmon resonance, shifting gradually from gold to silver and back indicates the formation of homogeneous alloy rather than segregated metal or core/shell type structure. The fungus on exposure to cadmium sulfate solution was found to yield cadmium sulfide quantum dots (5-20 nanometers) with hexagonal morphology. Long term incubation of the fungus with cadmium nitrate does not yield cadmium sulfide nanoparticles which indicate the action of sulfate reducing enzyme. Polyacrylamide gel electrophoresis of the extract led to four different protein bands. These proteins were extracted using dialysis and addition of ions to this solution does not yield cadmium sulfide which attests to the presence of some other factor. Addition of ATP and NADH restored the capability to produce quantum dots.

Nanoparticle Synthesis by Yeasts

Yeasts are most useful in the synthesis of semiconductor nanoparticles like cadmium sulfide, lead sulfide, antimony oxide, etc. Candida glubrata is the yeast which can intracellularly synthesize uniform spherically shaped peptide-bound CdS nano-crystals of size about 20 Å. They tend to form metal–thiolate complex with phytochelatins which neutralize of metal ions Schizosaccharomyces pombe can also synthesize CdS nano-crystals with hexagonal crystal structure of particle size 1-1.5 nm. Torulopsis sp. was the first yeast in which synthesis of face centered cubic structured PbS nano-crystals, showing semiconductor properties, which was intracellularly produced in the vacuoles having a dimension of 2–5 nm in spherical structure when incubated with Pb2+ reflects λ_{max} of 330 nm in UV–Vis spectrophotometer. Diode junction can be formed using these nanoparticles. S. cerevisiae (baker's yeast) can reduce Au^{+3} to give gold nanoparticles. Reduction happens in the peptidoglycan layer of the cell wall by the aldehyde group present in reducing sugars. Pichia jadinii is another yeast which can intracellularly synthesize the gold nanoparticles of different morphologies like spherical, triangular, hexagonal etc. of the size less than 100 nm in the cytoplasm of the cell within a day. Gold nanoparticles are also synthesized by the tropical marine yeast Yarrowia lipolytica by reducing the gold ions using pH control regulations. It synthesizes nanoparticles of various morphologies when subjected to different pH cultures. At pH 2.0 it produce hexagonal and triangular gold crystals because of the nucleation of the nanoparticles on the cell surfaces giving rise to golden color which falls in the visible range spectrum having wavelength 540 nm and at pH 7.0 and pH 9.0 gold nanoparticles gives pink and purple colors with an average size of ~ 15 nm. S. cerevisiae can also synthesize face-centered cubic unit cell antimony oxide (Sb_2O_3) nanoparticles with spherical morphology of size 2-10 nm at room temperature conditions. This was possibly due to the radial tautomerization of membrane-bound quinines or by membrane- bound or cytosolic pH-dependent oxidoreductases. Antimony oxides are an important ingredient for semiconductor industry. MKY3 is the only yeast so far discovered which is capable of extracellular silver nanoparticles synthesis of hexagonal crystal structure of size 2-5 nm in log phase growth of the yeast. These silver nanoparticles are used to make silver tolerant strain.

Nanoparticle	Bacteria	Fungi
Au	Rhodococcus sp.	Candida albican
	Shewanella oneidensis	Yarrowia lipolytica
	Plectonema boryanum UTEX 485	Neurospora crassa
	Escherichia coli	Phanerochaete chrysosporium
	Pseudomonas aeruginosa	Candida utilis
	Rhodopseudomonas capsulate	Neurospora crassa
	Brevibacterium casei	
	Ureibacillus thermosphaericus	
Ag	Bacillus licheniformis	Trichoderma viride
	Escherichia coli	Phaenerochaete chrysosporium

	Corynebacterium glutamicum	Aspergillus flavus
	Bacillus cereus	Aspergillus fumigatus
	Pseudomonas stutzeri	Verticillium sp.
		Fusarium oxysporum
		A. tubingensis
		Cladosporium cladosporioides
		Chrysosporium tropicum
		Phoma glomerata
Hg	Enterobacter sp	
Se	Shewanella sp	
CdTe	Escherichia coli	
Zn	Streptomyces sps	A. flavus
		A. terreus
		A. fumigatus
		A. tubingensis
Mg		A.tubingensis
		A. fumigatus
		Aspergillus brasiliensis

Ever-growing awareness about the necessity to turn towards environment-friendly approaches for materials synthesis to protect earth's environment has led to the development of eco-friendly, safe and green biological methods for nanoparticle production. Unlike other physical and chemical processes which involve the use expensive and sometimes hazardous chemicals and equipments, biological processes are cost-effective and eco-friendly. Hence, microbial synthesis of nanoparticles has emerged as an important part of nanotechnology. It is an interdisciplinary field which requires collaborations of physicists, chemists, biologists and engineers. Due to their vast diversity, bacteria, fungi and yeasts are able to produce a wide variety of nanoparticles and can act as biofactories for nanoparticles. However, a number of issues still need to be addressed both from nanotechnology and microbiology point of view before biosynthesis procedures can replace the traditional methods. The rate of production and monodispersity of the particles need to be improved and to achieve this microbial cultivation method and downstream processing have to be improved. The biochemical pathways involved in the synthesis of nanoparticles must be studied thoroughly and the specific genes and enzymes involved must be characterized. This will help us to have better control over the parameters that define the properties of a nanoparticle such as size, shape and monodispersity. Genetic engineering can also be used to make possible the use of organism whose growth process is known in detail and can be manipulated to our best interests. Future research on microbial synthesis of nanoparticle with unique properties is of great importance for applications in the field of medicine, agriculture and electronics.

Plant Metabolites

Nanotechnology is the term given to those areas of science and engineering where phenomena that take place at dimensions in the nanometre scale are utilised in the design, characterisation, production and application of materials, structures, devices and systems. Nano technology is one of the most rapidly progressing fields of technology and it has opened up numerous new frontiers of research for us. Its advent into the field targeted drug delivery, therapeutic actions and as bio sensors has captured the imagination of the scientific community and various methods are being devised to from new nanoparticles with more specifications, scientists are striving to come up with methods which let us control the shape, size, specificity and other characteristics of the particles more closely. One of the most useful and revolutionary technique coming up presently is synthesis of nanoparticles using plant extracts and their subsequent action. The formation of nano particles using plant extracts has a major edge over methods in terms of its interaction and effect on the environment; it is completely environmentally friendly and does not pose any threats even from its waste. The time required for the formation of particles is also within acceptable limits and with the ease of getting the requisite plants make it one of the best options available in this field to develop the particles. In this report we will expound various methods and the uses of manufacturing nano particles, which are fast becoming indispensable to us, using plants.

Significance of Plant Metabolytes and Uses

This report concerns synthesis of metal nanoparticles using plant metabolites. Even though nano particles can be developed using physiochemical techniques, they're lack of being environmentally benign causes a lot of problems. Especially, when there intended use is for the development of medicines. Environment factors are not the only reason biological synthesis is preferred, also because it can be used to produce large quantities of nanoparticles that are free of contamination and have a well-defined size and morphology. The use of plant metabolites to reduce metal ions has been known for a long time, although the nature of the reducing agents had been unknown for a long time. Processes for making nanoparticles using plant extracts are readily scalable and may be less expensive compared with the relatively expensive methods based on microbial processes or whole plants.

An important significance of plant extracts, in context of synthesizing nanoparticles, is that they act as both reducing and stabilizing agents. The nature of nanoparticle synthesized depends on the source of the plant extract. This aspect can also be utilized in making nanoparticle of preference. This happens because different sources of plant contain different concentrations and combinations of organic reducing agents.

Plants being used to reduce metal ions has been done for a long time, dating as back as early 1900s. But this practice was restricted to the use of whole plant extracts or plant tissues only. Compared to this the use of plant extracts to tct as both reducing agents and stabilizing agents.

Use of Plant Extracts in Nano Particle Synthesis

During the process of production of metal nanoparticles, the plant extract is simply mixed with a solution of metal salt at room temperature. It is a quick reaction and usually takes only minutes to complete. Nanoparticles of Gold, Silver and various other metals have been synthesized in the same way. Various plants' extract are used for synthesis of metal nano- particle, there are:

Neem (*Azadirachta indica*)   Aloe vera   Tea (*Camellia sinensis*);   *Catharanthus roseus*

Lemongrass (*Cymbopogon sp.*)   *Cinnamomum camphora*   *Datura metel*   Geranium leaf

Nanoparticle properties and production time depend on various characteristics of Plant extract, namely:

- its concentration,

- the concentration of the metal salt,

- the pH,

- temperature; and

- contact time

Advantages of using Plant extracts

1. The production of nano particles using the chemical methods has been raising concern among the environmentalists as they have an adverse affect on their ecology, hence the use of plant extracts for the formation of nano particles is being favoured due its salubrious nature towards the environment. Even in the industry it produces much less toxic waste.

2. The plants supplement both the reducing as well as stabilizing agents for the nano particles which otherwise have to be externally added in other methods.

3. The chemical method is being proven less economically beneficial as compared to the plant method as the maintenance cost is much less and the waste disposal requires less effort among other factors.

4. This method is even better than using the biological method as the maintenance of whole plant system is much less than a culture of bacteria which needs a myriad of phenomena to be taken care of.

5. Recent studies have shown that the therapeutic effects of plants , from which the nano particles are being derived, can also be imbued upon the particles hence providing us with perfect vehicles to the therapeutic materials to act upon the site of action as well as eliminating the need to artificially develop a drug for that particular ailment.

A few examples of production of metal nanoparticles using plant extracts are:

	Plant Species	Nanoparticles
1	Polyalthia longifolia	Silver Nanoparticles
2	Cassia auriculata	Gold Nanoparticles
3	Amarnath	Palladium Nano-particle
4	Neem leaf broth (Azadirachta indica)	Gold Nanoparticles
5	Jasminum grandiflorum	Silver Nanoparticles
6	Cymbopogon citrullus	Silver Nanoparticles

Silver Nanoparticles

- Leaf extract of Polyalthia longifolia was used synthesize silver nanoparticles (reported by Prasad and Elumalai, 2011). The average size of the particle hence formed was 58 nm. The reduction was ascribed to the phenolics, terpenoids, polysaccharides and flavones compounds present in the extract. Their bacterial activity peaked at 45 µg/mL (Huang etal., 2007).

- Stable size of 16-40 nm was acheived using geranium (Pelargoniumgraveolens) leaf extract. Geranoil, a natural monoterpene alcohol found in some plants, along with silver nitrate produced nanoparticles of range 1-10 nm.

- Sukirtha et al. (2011) synthesized silver nanoparticles using a leaf extract of Melia azedarach and showed them to be active against the HeLa cervical cancer cell line.

- Methanolic Extracts of Eucalyptus hybrida leaves have been reportedly used to synthesize silver nanoparticles. Flavonoid and terpenoid compounds present in the extract acted as stabilizing agents.

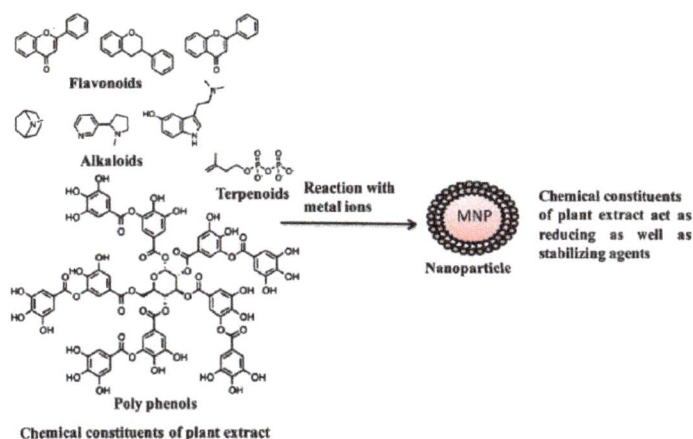

Chemical constituents of plant extract

Gold Nanoparticles

- Cassia auriculata has been used to synthesize spherical and triangular gold nano-particle of the size 15-25 nm. This process took only 10 minutes and was achieved at room temperature.

- Parida et al. (2011) reported the synthesis of gold nanoparticles mediated by an extract of Allium cepa. The particles had an average size of 100 nm and could be internalized by MCF-7 breast cancer cells via endocytosis (Parida et al., 2011)

- Leaf extracts of P. graveolens and Azadirachta indica are used to create nanoparticles of gold and silver, respectively. Aqueous Ag+ and Au+ ions are reduced to form bimetallic Ag and Au core-shell nano-particle.

Palladium Nanoparticles

- Amarnath et al. (2012) reported the antibacterial activity of palladium nanoparticles and their stabilization by chitosan and grape polyphenols. Palladium nanoparticles could be synthesized using coffee and tea extracts (Nadagouda and Varma, 2008). The nanoparticles were in the size range of 20–60 nm (Nadagouda and Varma, 2008).

- Lee et al. (2011) synthesized copper nanoparticles (40–100 nm) using magnolia leaf extract. These copper nanoparticles had antimicrobial activity against E. coli and were toxic to human adenocarcinomic alveolarbasal epithelial cells (A549 cells).

Application of Nano Particles

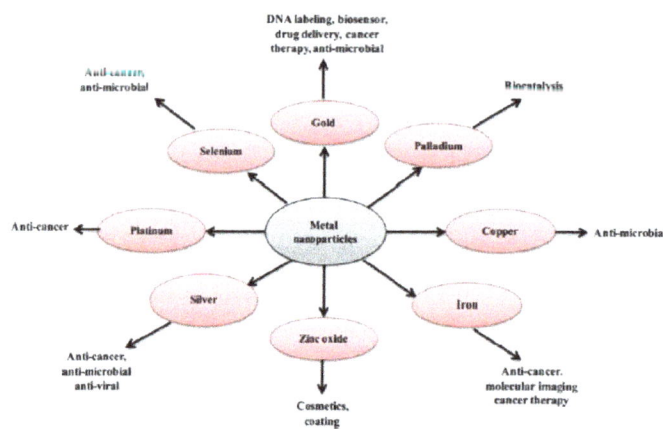

Various applications of nanoparticles synthesized from plant extracts.

Metal nanoparticle	Application	References
Silver	Anti-microbial, anti-cancer, anti-protozoal, Anti-fungal	Huang et al., 2007; Rajakumar and Abdul Rahuman, 2011; Sukirtha et al., 2011
Gold	Anti-microbial	Ghosh et al., 2011, 2012
Palladium	Anti-bacterial	Amarnath et al., 2012
Copper	Anti-microbial, anti-cancer	Lee et al., 2011
Selenium	Anti-cancer	Prasad et al., 2012

Nanoparticles are currently used in a variety of ways, including in vitro diagnostics, but their use in medicine is mostly on an experimental basis. The drugs based on nanoparticles have been claimed to have advantages over conventional drugs. These advantages include an extended half-life in vivo, longer circulation times and can convey a high concentration of drugs at specific targets. The size of nano-particle drug can be manipulated as discussed above. Due to high surface area of nano-particle drug loading becomes very high and can penetrate even the most impervious organs or tissues.

So far, nanoparticles have diverse in vitro diagnostic applications.

Applications

- Silver and Gold nanoparticles have found to have broad spectrum antimicrobial properties against human and animal pathogens. Silver is already being used in commercial medical and consumer products.

- Silver nanoparticles have larvicidal properties against malaria and filariasis vectors. They are also active against plasmodial pathogens and cancer. Antifungal effects have also been observed.

- Green tea plants can replace chemically hazardous compounds to a large extent.

- Nanoparticles are known to interfere with replication. This property has been used in anti cancer drugs using cisplatin.

- Nanoparticles have shown to reduce microbial loads in waste water treatment.

Protocol: Synthesis of Silver Nanoparticles using Acalypha Indica Leaf Extracts

Materials:

- The healthy leaves of A. indica were collected from campus of University of Madras, India. $AgNO_3$, MTT (methyl thiozolyl diphenyl-tetrazolium bromide) were purchased from Himedia Lab-oratories Pvt. Ltd., Mumbai, India. The bacterial cultures of E. coli(MTCC-443) and V. cholerae (MTCC-3904) were obtained from Microbial Type Culture Collection, Chandigarh, India.

- Preparation of plant extract Aqueous extract of A. indica was prepared using freshly collected leaves (10 g). They were surface cleaned with running tap water, followed by distilled water and boiled with 100 ml of distilled water at 60 °C for 5min. This extract was filtered through nylon mesh, followed by Millipore filter (0.45_m) and used for further experiments.

- Synthesis of silver nanoparticles For synthesis of silver nanoparticles, the Erlenmeyer flask containing 100 ml of $AgNO_3$ (1mM) was reacted with 12 ml of the aqueous extract of A. indica. This setup was incubated in dark (to minimize the photoactivation of silver nitrate), at 37 °C under static condition. A control setup was also maintained without A. Indica extract.

- Characterization of silver nanoparticles Synthesized silver nanoparticles was confirmed by sampling the reaction mixture at regular intervals and the absorption maxima wasscanned by UV– vis spectra, at the wavelength of 200–700nmin Beckman- DU 20 spectrophotometer. Further, the reaction mixture was subjected to centrifugation at 75,000×g for 30 min, resulting pellet was dissolved in deionized water and filtered through Millipore filter

(0.45_m). An aliquot of this filtrate containing silver nanoparticles was used for SEM, HR-TEM, XRD and EDS studies. For electron microscopic studies, 25_l of sample was sputter coated on copper stub and the images of nanoparticles were studied using SEM(JEOL, Model JFC-1600) and HRTEM(JEOL-3010). ForXRDstudies, dried nanoparticles were coated on XRD grid and the spectra was recorded by using PhilipsPW1830 X-ray generator operated at a voltage of 40 kV and a current of 30mA with Cu K_1 radiation. In addition presence of metals in the sample was analyzed by energy dispersive spectroscopy (EDS).

- Minimal inhibitory concentration of silver nanoparticles Minimal inhibitory concentrations (MICs) of AgNO3 and silver nanoparticles were determined by MTT assay by using 96-well microtitre plate . The mean of live cells of E. coli and V. Cholera was recorded using ELISA reader (Emax precision microplate reader). The MIC was determined based on different concentrations, where there was no increase in the OD595 and was zero.

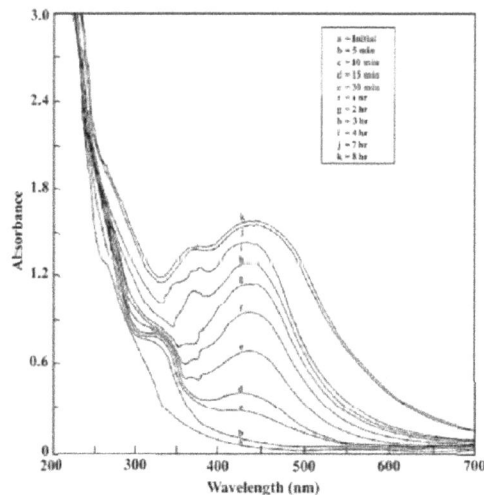

- UV−vis spectra of aqueous silver nitrate with A. indica leaf extract at different time intervals.

- Changes in membrane permeability of bacterial cells To study the membrane permeability of bacterial cells, the viable bacterial cultures of E. coli and V. cholerae in nutrient both were treated with synthesized silver nanoparticles. Ten milliliters of log phase cultures were centrifuged at 6000rpm for 10 min and the pellet was suspended in sterile distilled water. Five milliliters of this suspension was exposed to 100 ppb of silver nanoparticles and the conductance was recorded after incubation of 1, 3, 6 and 24 h using a conductivity meter (NAINA Model-NDC 732). The same procedure was adopted in the control experiments i.e. cultures treated with AgNO3.

- Determination of respiration activity of bacterial cells changes in the respiration of log phase cultures of E. coli and V. V.   cholerae in nutrient broth were studied using Biological Oxygen Monitor (YSI-Model-5300, USA). It provides a measure of oxygen consumption by the bacterial cultures. The changes in oxygen uptake among the untreated and silver nanoparticles treated cultures were recorded.

- EDAX analysis of silver nanoparticles showed characteristic peaks.

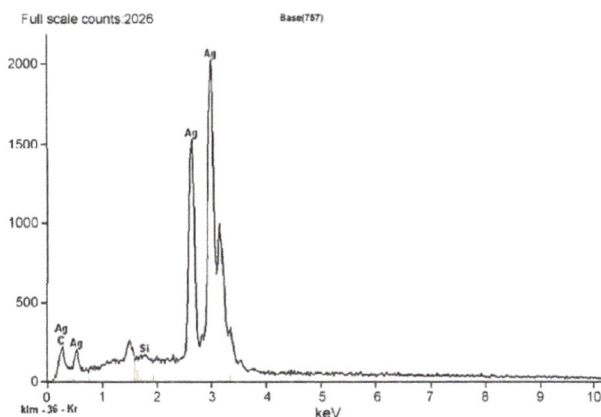

- Statistical analyses The data were subjected to One-way Analysis of Variance (ANOVA) to determine the significance of individual differences at $p < 0.05$ level. Significant means were compared by the Duncan's multiple range tests. All statistical analyses were carried out using SPSS statistical software package (SPSS, Version 10.0, Chicago, USA).

Concluding Remarks

The use of plant extracts for the synthesis of nanoparticles has proven to be inexpensive, easily scalable and most importantly environmentally benign. This aspect makes it indispensable for therapeutic applications. This method also ensures flexibility as manipulation regarding size and other properties is a characteristic implication of this procedure. Their antimicrobial properties have also been studies as discussed above. In vivo applications, however, are still under development.

Silver Nanoparticle

Silver nanoparticles are nanoparticles of silver of between 1 nm and 100 nm in size. While frequently described as being 'silver' some are composed of a large percentage of silver oxide due to their large ratio of surface-to-bulk silver atoms. Numerous shapes of nanoparticles can be constructed depending on the application at hand. Commonly used are spherical silver nanoparticles but diamond, octagonal and thin sheets are also popular.

Their extremely large surface area permits the coordination of a vast number of ligands. The properties of silver nanoparticles applicable to human treatments are under investigation in laboratory and animal studies, assessing potential efficacy, toxicity, and costs.

Synthetic Methods

Wet Chemistry

The most common methods for nanoparticle synthesis fall under the category of wet chemistry, or the nucleation of particles within a solution. This nucleation occurs when a silver ion complex, usually $AgNO_3$ or $AgClO_4$, is reduced to colloidal silver in the presence of a reducing agent. When the concentration increases enough, dissolved metallic silver ions bind together to form a stable surface. The surface is energetically unfavorable when the cluster is small, because the energy gained by decreasing the concentration of dissolved particles is not as high as the energy lost from creating a new surface. When the cluster reaches a certain size, known as the critical radius, it becomes energetically favorable, and thus stable enough to continue to grow. This nucleus then remains in the system and grows as more silver atoms diffuse through the solution and attach to the surface When the dissolved concentration of atomic silver decreases enough, it is no longer possible for enough atoms to bind together to form a stable nucleus. At this nucleation threshold, new nanoparticles stop being formed, and the remaining dissolved silver is absorbed by diffusion into the growing nanoparticles in the solution.

As the particles grow, other molecules in the solution diffuse and attach to the surface. This process stabilizes the surface energy of the particle and blocks new silver ions from reaching the surface. The attachment of these capping/stabilizing agents slows and eventually stops the growth of the particle. The most common capping ligands are trisodium citrate and polyvinylpyrrolidone (PVP), but many others are also used in varying conditions to synthesize particles with particular sizes, shapes, and surface properties.

There are many different wet synthesis methods, including the use of reducing sugars, citrate reduction, reduction via sodium borohydride, the silver mirror reaction, the polyol process, seed-mediated growth, and light-mediated growth. Each of these methods, or a combination of methods, will offer differing degrees of control over the size distribution as well as distributions of geometric arrangements of the nanoparticle.

A new, very promising wet-chemical technique was found by Elsupikhe et al. (2015). They have developed a green ultrasonically-assisted synthesis. Under ultrasound treatment, silver nanoparticles (AgNP) are synthesized with κ-carrageenan as a natural stabilizer. The reaction is performed at ambient temperature and produces silver nanoparticles with fcc crystal structure without impurities. The concentration of κ-carrageenan is used to influence particle size distribution of the AgNPs.

Monosaccharide Reduction

There are many ways silver nanoparticles can be synthesized; one method is through monosaccharides. This includes glucose, fructose, maltose, maltodextrin, etc. but not sucrose. It is also a simple method to reduce silver ions back to silver nanoparticles as it usually involves a one-step process,. There have been methods that indicated that these reducing sugars are essential to the formation of silver nanoparticles. Many studies indicated that this method of green synthesis, specifically using Cacumen platycladi extract, enabled the reduction of silver. Additionally, the size of the nanoparticle could be controlled depending on the concentration of the extract. The studies indicate that the higher concentrations correlated to an increased number of nanoparticles. Smaller nanoparticles were formed at high pH levels due to the concentration of the monosaccharides.

Another method of silver nanoparticle synthesis includes the use of reducing sugars with alkali starch and silver nitrate. The reducing sugars have free aldehyde and ketone groups, which enable them to be oxidized into gluconate. The monosaccharide must have a free ketone group because in order to act as a reducing agent it first undergoes tautomerization. In addition, if the aldehydes are bound, it will be stuck in cyclic form and cannot act as a reducing agent. For example, glucose has an aldehyde functional group that is able to reduce silver cations to silver atoms and is then oxidized to gluconic acid. The reaction for the sugars to be oxidized occurs in aqueous solutions. The capping agent is also not present when heated.

Citrate Reduction

An early, and very common, method for synthesizing silver nanoparticles is citrate reduction. This method was first recorded by M. C. Lea, who successfully produced a citrate-stabilized silver colloid in 1889. Citrate reduction involves the reduction of a silver source particle, usually $AgNO_3$ or $AgClO_4$, to colloidal silver using trisodium citrate, $Na_3C_6H_5O_7$. The synthesis is usually performed at an elevated temperature (~100 °C) to maximize the monodispersity (uniformity in both size and shape) of the particle. In this method, the citrate ion traditionally acts as both the reducing agent and the capping ligand, making it a useful process for AgNP production due to its relative ease and short reaction time. However, the silver particles formed may exhibit broad size distributions and form several different particle geometries simultaneously. The addition of stronger reducing agents to the reaction is often used to synthesize particles of a more uniform size and shape.

Reduction via Sodium Borohydride

The synthesis of silver nanoparticles by sodium borohydride ($NaBH_4$) reduction occurs by the following reaction:

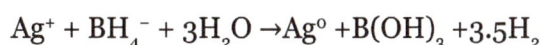

$$Ag^+ + BH_4^- + 3H_2O \rightarrow Ag^0 + B(OH)_3 + 3.5H_2$$

The reduced metal atoms will form nanoparticle nuclei. Overall, this process is similar to the above reduction method using citrate. The benefit of using sodium borohydride is increased monodispersity of the final particle population. The reason for the increased monodispersity when using $NaBH_4$ is that it is a stronger reducing agent than citrate. The impact of reducing agent strength can be seen by inspecting a LaMer diagram which describes the nucleation and growth of nanoparticles.

When silver nitrate ($AgNO_3$) is reduced by a weak reducing agent like citrate, the reduction rate is lower which means that new nuclei are forming and old nuclei are growing concurrently. This is the reason that the citrate reaction has low monodispersity. Because $NaBH_4$ is a much stronger reducing agent, the concentration of silver nitrate is reduced rapidly which shortens the time during which new nuclei form and grow concurrently yielding a monodispersed population of silver nanoparticles.

Particles formed by reduction must have their surfaces stabilized to prevent undesirable particle agglomeration (when multiple particles bond together), growth, or coarsening. The driving force for these phenomena is the minimization of surface energy (nanoparticles have a large surface to volume ratio). This tendency to reduce surface energy in the system can be counteracted by adding species which will adsorb to the surface of the nanoparticles and lowers the activity of the particle surface thus preventing particle agglomeration according to the DLVO theory and preventing growth by occupying attachment sites for metal atoms. Chemical species that adsorb to the surface of nanoparticles are called ligands. Some of these surface stabilizing species are: $NaBH_4$ in large amounts, poly(vinyl pyrrolidone) (PVP), sodium dodecyl sulfate (SDS), and/or dodecane thiol.

Once the particles have been formed in solution they must be separated and collected. There are several general methods to remove nanoparticles from solution, including evaporating the solvent phase or the addition of chemicals to the solution that lower the solubility of the nanoparticles in the solution. Both methods force the precipitation of the nanoparticles.

Polyol Process

The polyol process is a particularly useful method because it yields a high degree of control over both the size and geometry of the resulting nanoparticles. In general, the polyol synthesis begins with the heating of a polyol compound such as ethylene glycol, 1,5-pentanediol, or 1,2-propylene glycol7. An Ag^+ species and a capping agent are added (although the polyol itself is also often the capping agent). The Ag^+ species is then reduced by the polyol to colloidal nanoparticles. The polyol process is highly sensitive to reaction conditions such as temperature, chemical environment, and concentration of substrates. Therefore, by changing these variables, various sizes and geometries can be selected for such as quasi-spheres, pyramids, spheres, and wires. Further study has examined the mechanism for this process as well as resulting geometries under various reaction conditions in greater detail.

Seed-mediated Growth

Seed-mediated growth is a synthetic method in which small, stable nuclei are grown in a separate chemical environment to a desired size and shape. Seed-mediated methods consist of two different stages: nucleation and growth. Variation of certain factors in the synthesis (e.g. ligand, nucleation time, reducing agent, etc.), can control the final size and shape of nanoparticles, making seed-mediated growth a popular synthetic approach to controlling morphology of nanoparticles.

The nucleation stage of seed-mediated growth consists of the reduction of metal ions in a precursor to metal atoms. In order to control the size distribution of the seeds, the period of nucleation should be made short for monodispersity. The LaMer model illustrates this concept. Seeds typically consist small nanoparticles, stabilized by a ligand. Ligands are small, usually organic molecules that bind to the surface of particles, preventing seeds from further growth. Ligands are necessary as they increase the energy barrier of coagulation, preventing agglomeration. The balance between attractive and repulsive forces within colloidal solutions can be modeled by DLVO theory. Ligand binding affinity, and selectivity can be used to control shape and growth. For seed synthesis, a ligand with medium to low binding affinity should be chosen as to allow for exchange during growth phase.

The growth of nanoseeds involves placing the seeds into a growth solution. The growth solution requires a low concentration of a metal precursor, ligands that will readily exchange with preexisting seed ligands, and a weak or very low concentration of reducing agent. The reducing agent must not be strong enough to reduce metal precursor in the growth solution in the absence of seeds. Otherwise, the growth solution will form new nucleation sites instead of growing on preexisting ones (seeds). Growth is the result of the competition between surface energy (which increases unfavorably with growth) and bulk energy (which decreases favorably with growth). The balance between the energetics of growth and dissolution is the reason for uniform growth only on preexisting seeds (and no new nucleation). Growth occurs by the addition of metal atoms from the growth solution to the seeds, and ligand exchange between the growth ligands (which have a higher bonding affinity) and the seed ligands.

Range and direction of growth can be controlled by nanoseed, concentration of metal precursor, ligand, and reaction conditions (heat, pressure, etc.). Controlling stoichiometric conditions of growth solution controls ultimate size of particle. For example, a low concentration of metal seeds to metal precursor in the growth solution will produce larger particles. Capping agent has been shown to control direction of growth and thereby shape. Ligands can have varying affinities for binding across a particle. Differential binding within a particle can result in dissimilar growth across particle. This produces anisotropic particles with nonspherical shapes including prisms, cubes, and rods.

Light-mediated Growth

Light-mediated syntheses have also been explored where light can promote formation of various silver nanoparticle morphologies.

Silver Mirror Reaction

The silver mirror reaction involves the conversion of silver nitrate to Ag(NH3)OH. Ag(NH3)OH is subsequently reduced into colloidal silver using an aldehyde containing molecule such as a sugar. The silver mirror reaction is as follows:

$$2(Ag(NH_3)_2)^+ + RCHO + 2OH^- \rightarrow RCOOH + 2Ag + 4NH_3.$$

The size and shape of the nanoparticles produced are difficult to control and often have wide distributions. However, this method is often used to apply thin coatings of silver particles onto surfaces and further study into producing more uniformly sized nanoparticles is being done.

Ion Implantation

Ion implantation has been used to create silver nanoparticles embedded in glass, polyurethane, silicone, polyethylene, and poly(methyl methacrylate). Particles are embedded in the substrate by means of bombardment at high accelerating voltages. At a fixed current density of the ion beam up to a certain value, the size of the embedded silver nanoparticles has been found to be monodisperse within the population, after which only an increase in the ion concentration is observed. A further increase in the ion beam dose has been found to reduce both the nanoparticle size and density in the target substrate, whereas an ion beam operating at a high accelerating voltage with a gradually increasing current density has been found to result in a gradual increase in the nanoparticle size. There are a few competing mechanisms which may result in the decrease in nanoparticle size; destruction of NPs upon collision, sputtering of the sample surface, particle fusion upon heating and dissociation.

The formation of embedded nanoparticles is complex, and all of the controlling parameters and factors have not yet been investigated. Computer simulation is still difficult as it involves processes of diffusion and clustering, however it can be broken down into a few different sub-processes such as implantation, diffusion, and growth. Upon implantation, silver ions will reach different depths within the substrate which approaches a Gaussian distribution with the mean centered at X depth. High temperature conditions during the initial stages of implantation will increase the impurity diffusion in the substrate and as a result limit the impinging ion saturation, which is required for nanoparticle nucleation. Both the implant temperature and ion beam current density are crucial to control in order to obtain a monodisperse nanoparticle size and depth distribution. A low current density may be used to counter the thermal agitation from the ion beam and a buildup of surface charge. After implantation on the surface, the beam currents may be raised as the surface conductivity will increase. The rate at which impurities diffuse drops quickly after the formation of the nanoparticles, which act as a mobile ion trap. This suggests that the beginning of the implantation process is critical for control of the spacing and depth of the resulting nanoparticles, as well as control of the substrate temperature and ion beam density. The presence and nature of these particles can be analyzed using numerous spectroscopy and microscopy instruments. Nanoparticles synthesized in the substrate exhibit surface plasmon resonances as evidenced by characteristic absorption bands; these features undergo spectral shifts depending on the nanoparticle size and surface asperities, however the optical properties also strongly depend on the substrate material of the composite.

Biological Synthesis

The biological synthesis of nanoparticles has provided a means for improved techniques compared to the traditional methods that call for the use of harmful reducing agents like sodium borohydride. Many of these methods could improve their environmental footprint by replacing these relatively strong reducing agents. The problems with the chemical production of silver nanoparticles is usually involves high cost and the longevity of the particles is short lived due to aggregation. The harshness of standard chemical methods has sparked the use of using biological organisms to reduce silver ions in solution into colloidal nanoparticles.

In addition, precise control over shape and size is vital during nanoparticle synthesis since the NPs therapeutic properties are intimately dependent on such factors. Hence, the primary focus of research in biogenic synthesis is in developing methods that consistently reproduce NPs with precise properties.

Fungi and Bacteria

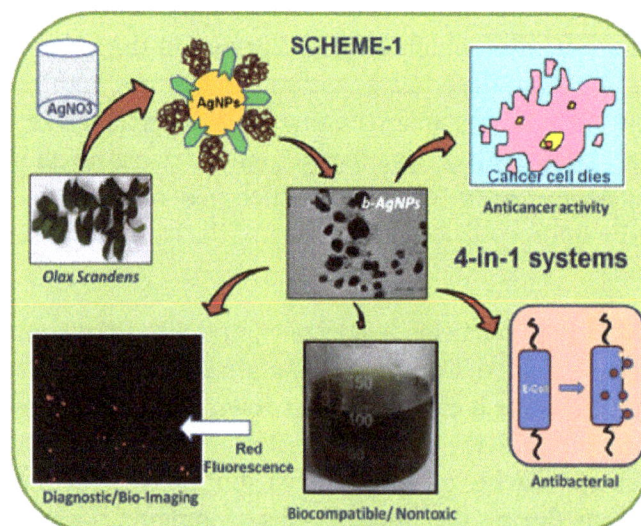

A general representation of the synthesis and applications of biogenically synthesized silver nanoparticles using plant extract.

Bacterial and fungal synthesis of nanoparticles is practical because bacteria and fungi are easy to handle and can be modified genetically with ease. This provides a means to develop biomolecules that can synthesize AgNPs of varying shapes and sizes in high yield, which is at the forefront of current challenges in nanoparticle synthesis. Fungal strains such as Verticillium and bacterial strains such as K. pneumoniae can be used in the synthesis of silver nanoparticles. When the fungus/bacteria is added to solution, protein biomass is released into the solution. Electron donating residues such as tryptophan and tyrosine reduce silver ions in solution contributed by silver nitrate. These methods have been found to effectively create stable monodisperse nanoparticles without the use of harmful reducing agents.

A method has been found of reducing silver ions by the introduction of the fungus *Fusarium oxysporum*. The nanoparticles formed in this method have a size range between 5 and 15 nm and consist of silver hydrosol. The reduction of the silver nanoparticles is thought to come from an enzymatic process and silver nanoparticles produced are extremely stable due to interactions with proteins that are excreted by the fungi.

Bacterium found in silver mines, Pseudomonas stutzeri AG259, were able to construct silver particles in the shapes of triangles and hexagons. The size of these nanoparticles had a large range in size and some of them reached sizes larger than the usual nanoscale with a size of 200 nm. The silver nanoparticles were found in the organic matrix of the bacteria.

Lactic acid producing bacteria have been used to produce silver nanoparticles. The bacteria *Lactobacillus* spp., *Pediococcus pentosaceus, Enteroccus faeciumI*, and *Lactococcus garvieae* have been found to be able to reduce silver ions into silver nanoparticles. The production of the nanoparticles takes place in the cell from the interactions between the silver ions and the organic compounds of the cell. It was found that the bacterium *Lactobacillus fermentum* created the smallest silver nanoparticles with an average size of 11.2 nm. It was also found that this bacterium produced the nanoparticles with the smallest size distribution and the nanoparticles were found mostly on the outside of the cells. It was also found that there was an increase in the pH increased the rate of which the nanoparticles were produced and the amount of particles produced.

Plants

The reduction of silver ions into silver nanoparticles has also been achieved using geranium leaves. It has been found that adding geranium leaf extract to silver nitrate solutions causes their silver ions to be quickly reduced and that the nanoparticles produced are particularly stable. The silver nanoparticles produced in solution had a size range between 16 and 40 nm.

In another study different plant leaf extracts were used to reduce silver ions. It was found that out of Camellia sinensis (green tea), pine, persimmon, ginko, magnolia, and platanus that the magnolia leaf extract was the best at creating silver nanoparticles. This method created particles with a disperse size range of 15 to 500 nm, but it was also found that the particle size could be controlled by varying the reaction temperature. The speed at which the ions were reduced by the magnolia leaf extract was comparable to those of using chemicals to reduce.

The use of plants, microbes, and fungi in the production of silver nanoparticles is leading the way to more environmentally sound production of silver nanoparticles.

A green method is available for synthesizing silver nanoparticles using Amaranthus gangeticus Linn leaf extract.

Products and Functionalization

Synthetic protocols for silver nanoparticle production can be modified to produce silver nanoparticles with non-spherical geometries and also to functionalize nanoparticles with different materials, such as silica. Creating silver nanoparticles of different shapes and surface coatings allows for greater control over their size-specific properties.

Anisotropic Structures

Silver nanoparticles can be synthesized in a variety of non-spherical (anisotropic) shapes. Because silver, like other noble metals, exhibits a size and shape dependent optical effect known as localized surface plasmon resonance (LSPR) at the nanoscale, the ability to synthesize Ag nanoparticles in different shapes vastly increases the ability to tune their optical behavior. For example, the wavelength at which LSPR occurs for a nanoparticle of one morphology (e.g. a sphere) will be different if that sphere is changed into a different shape. This shape dependence allows a silver nanoparticle to experience optical enhancement at a range of different wavelengths, even by keeping the size relatively constant, just by changing its shape. The applications of this shape-exploited expansion of optical behavior range from developing more sensitive biosensors to increasing the longevity of textiles.

Triangular Nanoprisms

Triangular shaped nanoparticles are a canonical type of anisotropic morphology studied for both gold and silver.

Though many different techniques for silver nanoprism synthesis exist, several methods employ a seed-mediated approach, which involves first synthesizing small (3-5 nm diameter) silver nanoparticles that offer a template for shape-directed growth into triangular nanostructures.

The silver seeds are synthesized by mixing silver nitrate and sodium citrate in aqueous solution and then rapidly adding sodium borohydride. Additional silver nitrate is added to the seed solution at low temperature, and the prisms are grown by slowly reducing the excess silver nitrate using ascorbic acid.

With the seed-mediated approach to silver nanoprism synthesis, selectivity of one shape over another can in part be controlled by the capping ligand. Using essentially the same procedure above but changing citrate to poly (vinyl pyrrolidone) (PVP) yields cube and rod-shaped nanostructures instead of triangular nanoprisms.

In addition to the seed mediated technique, silver nanoprisms can also be synthesized using a photo-mediated approach, in which preexisting spherical silver nanoparticles are transformed into triangular nanoprisms simply by exposing the reaction mixture to high intensities of light.

Nanocubes

Silver nanocubes can be synthesized using ethylene glycol as a reducing agent and PVP as a capping agent, in a polyol synthesis reaction (vide supra). A typical synthesis using these reagents involves adding fresh silver nitrate and PVP to a solution of ethylene glycol heated at 140 °C.

This procedure can actually be modified to produce another anisotropic silver nanostructure, nanowires, by just allowing the silver nitrate solution to age before using it in the synthesis. By allowing the silver nitrate solution to age, the initial nanostructure formed during the synthesis is slightly different than that obtained with fresh silver nitrate, which influences the growth process, and therefore, the morphology of the final product.

Coating with Silica

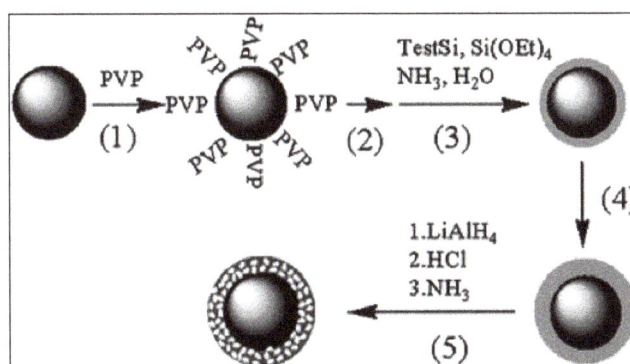

General procedure for coating colloid particles in silica. First PVP is absorbed onto the colloidal surface. These particles are put into a solution of ammonia in ethanol. the particle then begins to grow by addition of Si(OET4).

In this method, polyvinylpyrrolidone (PVP) is dissolved in water by sonication and mixed with silver colloid particles. Active stirring ensures the PVP has adsorbed to the nanoparticle surface. Centrifuging separates the PVP coated nanoparticles which are then transferred to a solution of ethanol to be centrifuged further and placed in a solution of ammonia, ethanol and $Si(OEt_4)$ (TES). Stirring for twelve hours results in the silica shell being formed consisting of a surrounding layer of silicon oxide with an ether linkage available to add functionality. Varying the amount of TES allows for different thicknesses of shells formed. This technique is popular due to the ability to add a variety of functionality to the exposed silica surface.

Use

Catalysis

Using silver nanoparticles for catalysis has been gaining attention in recent years. Although the most common applications are for medicinal or antibacterial purposes, silver nanoparticles have been demonstrated to show catalytic redox properties for dyes, benzene, carbon monoxide, and likely other compounds.

NOTE: This paragraph is a general description of nanoparticle properties for catalysis; it is not exclusive to silver nanoparticles. The size of a nanoparticle greatly determines the properties that it exhibits due to various quantum effects. Additionally, the chemical environment of the nanoparticle plays a large role on the catalytic properties. With this in mind, it is important to note that heterogeneous catalysis takes place by adsorption of the reactant species to the catalytic substrate. When polymers, complex ligands, or surfactants are used to prevent coalescence of the nanoparticles, the catalytic ability is frequently hindered due to reduced adsorption ability. However, these compounds can also be used in such a way that the chemical environment enhances the catalytic ability.

Supported on Silica Spheres – Reduction of Dyes

Silver nanoparticles have been synthesized on a support of inert silica spheres. The support plays virtually no role in the catalytic ability and serves as a method of preventing coalescence of the silver nanoparticles in colloidal solution. Thus, the silver nanoparticles were stabilized and it was possible to demonstrate the ability of them to serve as an electron relay for the reduction of dyes by sodium borohydride. Without the silver nanoparticle catalyst, virtually no reaction occurs between sodium borohydride and the various dyes: methylene blue, eosin, and rose bengal.

Mesoporous Aerogel – Selective Oxidation of Benzene

Silver nanoparticles supported on aerogel are advantageous due to the higher number of active sites. The highest selectivity for oxidation of benzene to phenol was observed at low weight percent of silver in the aerogel matrix (1% Ag). This better selectivity is believed to be a result of the higher monodispersity within the aerogel matrix of the 1% Ag sample. Each weight percent solution formed different sized particles with a different width of size range.

Silver Alloy – Synergistic Oxidation of Carbon Monoxide

Au-Ag alloy nanoparticles have been shown to have a synergistic effect on the oxidation of carbon monoxide (CO). On its own, each pure-metal nanoparticle shows very poor catalytic activity for CO oxidation; together, the catalytic properties are greatly enhanced. It is proposed that the gold acts as a strong binding agent for the oxygen atom and the silver serves as a strong oxidizing catalyst, although the exact mechanism is still not completely understood. When synthesized in an Au/Ag ratio from 3:1 to 10:1, the alloyed nanoparticles showed complete conversion when 1% CO was fed in air at ambient temperature. Interestingly, the size of the alloyed particles did not play a big role in the catalytic ability. It is well known that gold nanoparticles only show catalytic properties for CO when they are ~3 nm in size, but alloyed particles up to 30 nm demonstrated excellent catalytic activity – catalytic activity better than that of gold nanoparticles on active support such as TiO_2, Fe_2O_3, etc.

Light-enhanced

Plasmonic effects have been studied quite extensively. Until recently, there have not been studies investigating the oxidative catalytic enhancement of a nanostructure via excitation of its surface plasmon resonance. The defining feature for enhancing the oxidative catalytic ability has been identified as the ability to convert a beam of light into the form of energetic electrons that can be transferred to adsorbed molecules. The implication of such a feature is that photochemical reactions can be driven by low-intensity continuous light can be coupled with thermal energy.

The coupling of low-intensity continuous light and thermal energy has been performed with silver nanocubes. The important feature of silver nanostructures that are enabling for photocatalysis is their nature to create resonant surface plasmons from light in the visible range.

The addition of light enhancement enabled the particles to perform to the same degree as particles that were heated up to 40 K greater. This is a profound finding when noting that a reduction in temperature of 25 K can increase the catalyst lifetime by nearly tenfold, when comparing the photothermal and thermal process.

Biological Research

Researchers have explored the use of silver nanoparticles as carriers for delivering various payloads such as small drug molecules or large biomolecules to specific targets. Once the AgNP has had sufficient time to reach its target, release of the payload could potentially be triggered by an internal or external stimulus. The targeting and accumulation of nanoparticles may provide high payload concentrations at specific target sites and could minimize side effects.

Chemotherapy

The introduction of nanotechnology into medicine is expected to advance diagnostic cancer imaging and the standards for therapeutic drug design. Nanotechnology may uncover insight about the structure, function and organizational level of the biosystem at the nanoscale.

Silver nanoparticles can undergo coating techniques that offer a uniform functionalized surface to which substrates can be added. When the nanoparticle is coated, for example, in silica the surface exists as silicic acid. Substrates can thus be added through stable ether and ester linkages that are not degraded immediately by natural metabolic enzymes. Recent chemotherapeutic applications have designed anti cancer drugs with a photo cleavable linker, such as an ortho-nitrobenzyl bridge, attaching it to the substrate on the nanoparticle surface. The low toxicity nanoparticle complex can remain viable under metabolic attack for the time necessary to be distributed throughout the bodies systems. If a cancerous tumor is being targeted for treatment, ultraviolet light can be introduced over the tumor region. The electromagnetic energy of the light causes the photo responsive linker to break between the drug and the nanoparticle substrate. The drug is now cleaved and released in an unaltered active form to act on the cancerous tumor cells. Advantages anticipated for this method is that the drug is transported without highly toxic compounds, the drug is released without harmful radiation or relying on a specific chemical reaction to occur and the drug can be selectively released at a target tissue.

A second approach is to attach a chemotherapeutic drug directly to the functionalized surface of the silver nanoparticle combined with a nucelophilic species to undergo a displacement reaction. For example, once the nanoparticle drug complex enters or is in the vicinity of the target tissue or cells, a glutathione monoester can be administered to the site. The nucleophilic ester oxygen will attach to the functionalized surface of the nanoparticle through a new ester linkage while the drug is released to its surroundings. The drug is now active and can exert its biological function on the cells immediate to its surroundings limiting non-desirable interactions with other tissues.

Multiple Drug Resistance

A major cause for the ineffectiveness of current chemotherapy treatments is multiple drug resistance which can arise from several mechanisms.

Nanoparticles can provide a means to overcome MDR. In general, when using a targeting agent to deliver nanocarriers to cancer cells, it is imperative that the agent binds with high selectivity to molecules that are uniquely expressed on the cell surface. Hence NPs can be designed with proteins that specifically detect drug resistant cells with overexpressed transporter proteins on their surface. A pitfall of the commonly used nano-drug delivery systems is that free drugs that are released from the nanocarriers into the cytosol get exposed to the MDR transporters once again, and are exported. To solve this, 8 nm nano crystalline silver particles were modified by the addition of trans-activating transcriptional activator (TAT), derived from the HIV-1 virus, which acts as a cell penetrating peptide (CPP). Generally, AgNP effectiveness is limited due to the lack of efficient cellular uptake; however, CPP-modification has become one of the most efficient methods for improving intracellular delivery of nanoparticles. Once ingested, the export of the AgNP is prevented based on a size exclusion. The concept is simple: the nanoparticles are too large to be effluxed by the MDR transporters, because the efflux function is strictly subjected to the size of its substrates, which is generally limited to a range of 300-2000 Da. Thereby the nanoparticulates remain insusceptible to the efflux, providing a means to accumulate in high concentrations.

Antimicrobial

Introduction of silver into bacterial cells induces a high degree of structural and morphological changes, which can lead to cell death. As the silver nano particles come in contact with the bacteria, they adhere to the cell wall and cell membrane. Once bound, some of the silver passes through to the inside, and interacts with phosphate-containing compounds like DNA and RNA, while another portion adheres to the sulphur-containing proteins on the membrane. The silver-sulphur interactions at the membrane cause the cell wall to undergo structural changes, like the formation of pits and pores. Through these pores, cellular components are released into the extracellular fluid, simply due to the osmotic difference. Within the cell, the integration of silver creates a low molecular weight region where the DNA then condenses. Having DNA in a condensed state inhibits the cell's replication proteins contact with the DNA. Thus the introduction of silver nanoparticles inhibits replication and is sufficient to cause the death of the cell. Further increasing their effect, when silver comes in contact with fluids, it tends to ionize which increases the nanoparticles bactericidal activity. This has been correlated to the suppression of enzymes and inhibited expression of proteins that relate to the cell's ability to produce ATP.

Although it varies for every type of cell proposed, as their cell membrane composition varies greatly, It has been seen that in general, silver nano particles with an average size of 10 nm or less show electronic effects that greatly increase their bactericidal activity. This could also be partly due to the fact that as particle size decreases, reactivity increases due to the surface area to volume ratio increasing.

It has been noted that the introduction of silver nano particles has shown to have synergistic activity with common antibiotics already used today, such as; penicillin G, ampicillin, erythromycin, clindamycin, and vancomycin against E. coli and S. aureus. In medical equipment, it has been shown that silver nano particles drastically lower the bacterial count on devices used. However, the problem arises when the procedure is over and a new one must be done. In the process of washing the instruments a large portion of the silver nano particles become less effective due to the loss of silver ions. They are more commonly used in skin grafts for burn victims as the silver nano particles embedded with the graft provide better antimicrobial activity and result in significantly less scarring of the victim. They also show promising application as water treatment method to form clean potable water.

Silver nanoparticles can prevent bacteria from growing on or adhering to the surface. This can be especially useful in surgical settings where all surfaces in contact with the patient must be sterile.

silver in solution, determined that although initially silver ions were 18 times more likely to inhibit the photosynthesis of an algae, Chlamydomanas reinhardtii, but after 2 hours of incubation it was revealed that the algae containing silver nanoparticles were more toxic than just silver ions alone. Furthermore, there are studies that suggest that silver nanoparticles induce toxicity independent of free silver ions. For example, Asharani *et al.* compared phenotypic defects observed in zebrafish treated with silver nanoparticles and silver ions and determined that the phenotypic defects observed with silver nanoparticle treatment was not observed with silver ion-treated embryos, suggesting that the toxicity of silver nanoparticles are independent of silver ions.

Protein channels and nuclear membrane pores can often be in the size range of 9 nm to 10 nm in diameter. Small silver nanoparticles constructed of this size have the ability to not only pass through the membrane to interact with internal structures but also to be become lodged within the membrane. Silver nanoparticle depositions in the membrane can impact regulation of solutes, exchange of proteins and cell recognition. Exposure to silver nanoparticles has been associated with "inflammatory, oxidative, genotoxic, and cytotoxic consequences"; the silver particulates primarily accumulate in the liver. but have also been shown to be toxic in other organs including the brain. Nano-silver applied to tissue-cultured human cells leads to the formation of free radicals, raising concerns of potential health risks.

- Allergic reaction: There have been several studies conducted that show a precedence for allerginicity of silver nanoparticles.

- Argyria and staining: Ingested silver or silver compounds, including colloidal silver, can cause a condition called argyria, a discoloration of the skin and organs.In 2006, there was a case study of a 17-year-old man, who sustained burns to 30% of his body, and experienced a temporary bluish-grey hue after several days of treatment with Acticoat, a brand of wound dressing containing silver nanoparticles. Argyria is the deposition of silver in deep tissues, a condition that cannot happen on a temporary basis, raising the question of whether the cause of the man's discoloration was argyria or even a result of the silver treatment. Silver dressings are known to cause a "transient discoloration" that dissipates in 2–14 days, but not a permanent discoloration.

- Silzone heart valve: St. Jude Medical released a mechanical heart valve with a silver coated sewing cuff (coated using ion beam-assisted deposition) in 1997. The valve was designed to reduce the instances of endocarditis. The valve was approved for sale in Canada, Europe, the United States, and most other markets around the world. In a post-commercialization study, researchers showed that the valve prevented tissue ingrowth, created paravalvular leakage, valve loosening, and in the worst cases explantation. After 3 years on the market and 36,000 implants, St. Jude discontinued and voluntarily recalled the valve.

Colloidal Gold

Colloidal gold is a sol or colloidal suspension of nanoparticles of gold in a fluid, usually water. The liquid is usually either an intense red colour (for particles less than 100 nm) or blue/purple (for larger particles). Due to the unique optical, electronic, and molecular-recognition properties of gold nanoparticles, they are the subject of substantial research, with applications in a wide variety of areas, including electron microscopy, electronics, nanotechnology, and materials science.

82 Nanobiotechnology: An Introduction

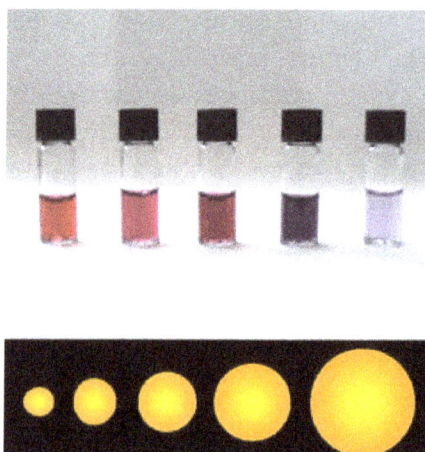

Suspensions of gold nanoparticles of various sizes. The size difference causes the difference in colors.

The properties of colloidal gold nanoparticles, and thus their applications, depend strongly upon their size and shape. For example, rodlike particles have both transverse and longitudinal absorption peak, and anisotropy of the shape affects their self-assembly.

History

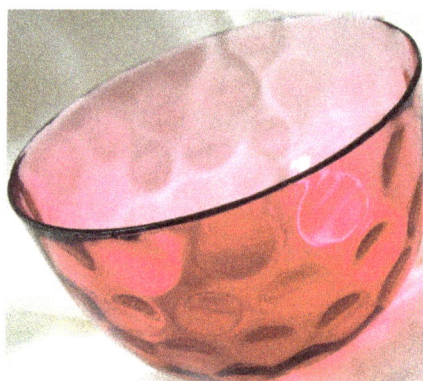

This cranberry glass bowl was made by adding a gold salt (probably gold chloride) to molten glass.

Known, or at least used (perhaps proceeding by accident without much understanding of the process) since ancient times, the synthesis of colloidal gold was crucial to the 4th-century Lycurgus Cup, which changes color depending on the location of light source. Later it was used as a method of staining glass.

During the Middle Ages, soluble gold, a solution containing gold salt, had a reputation for its curative property for various diseases. In 1618, Francis Anthony, a philosopher and member of the medical profession, published a book called *Panacea Aurea, sive tractatus duo de ipsius Auro Potabili* (Latin: gold potion, or two treatments of potable gold). The book introduces information on the formation of colloidal gold and its medical uses. About half a century later, English botanist Nicholas Culpepper published book in 1656, *Treatise of Aurum Potabile*, solely discussing the medical uses of colloidal gold.

In 1676, Johann Kunckel, a German chemist, published a book on the manufacture of stained glass. In his book *Valuable Observations or Remarks About the Fixed and Volatile Salts-Auro and Argento Potabile, Spiritu Mundi and the Like*, Kunckel assumed that the slight pink color of Aurum Potabile came from small particles of metallic gold, not visible to human eyes. In 1842, John Herschel invented a photographic process called chrysotype that used colloidal gold to record images on paper.

Modern scientific evaluation of colloidal gold did not begin until Michael Faraday's work in the 1850s. In 1856, in a basement laboratory of Royal Institution, Faraday accidentally created a ruby red solution while mounting pieces of gold leaf onto microscope slides. Since he was already interested in the properties of light and matter, Faraday further investigated the optical properties of the colloidal gold. He prepared the first pure sample of colloidal gold, which he called 'activated gold', in 1857. He used phosphorus to reduce a solution of gold chloride. The colloidal gold Faraday made 150 years ago is still optically active. For a long time, the composition of the 'ruby' gold was unclear. Several chemists suspected it to be a gold tin compound, due to its preparation. Faraday recognized that the color was actually due to the miniature size of the gold particles. He noted the light scattering properties of suspended gold microparticles, which is now called Faraday-Tyndall effect.

In 1898, Richard Adolf Zsigmondy prepared the first colloidal gold in diluted solution. Apart from Zsigmondy, Theodor Svedberg, who invented ultracentrifugation, and Gustav Mie, who provided the theory for scattering and absorption by spherical particles, were also interested in the synthesis and properties of colloidal gold.

With advances in various analytical technologies in the 20th century, studies on gold nanoparticles has accelerated. Advanced microscopy methods, such as atomic force microscopy and electron microscopy, have contributed the most to nanoparticle research. Due to their comparably easy synthesis and high stability, various gold particles have been studied for their practical uses. Different types of gold nanoparticle are already used in many industries, such as medicine and electronics. For example, several FDA-approved nanoparticles are currently used in drug delivery.

Physical Properties

Optical

Colloidal gold has been used by artists for centuries because of the nanoparticle's interactions with visible light. Gold nanoparticles absorb and scatter light with incredible efficiency. Ranging from vibrant reds to blues to black and finally to clear and colorless, colloidal gold has the ability to exhibit a wide range of colors depending on particle size, shape, local refractive index, and aggregation state. These colors occur because of a phenomenon called Localized Surface Plasmon Resonance (LSPR), in which conduction electrons on the surface of the nanoparticle oscillate in resonance with incident light.

Effect of Size

As a general rule, the wavelength of light absorbed increases as a function of increasing nano particle size. For example, pseudo-spherical gold nanoparticles with diameters ~ 30 nm have a peak LSPR absorption at ~530 nm.

Effect of Local Refractive Index

Changes in the apparent color of a gold nanoparticle solution can also be caused by the environment in which the colloidal gold is suspended The optical properties of gold nanoparticles depends on the refractive index near the nanoparticle surface, therefore both the molecules directly attached to the nanoparticle surface (i.e. nanoparticle ligands) and/or the nanoparticle solvent both may influence observed optical features. As the refractive index near the gold surface increases, the NP LSPR will shift to longer wavelengths In addition to solvent environment, the extinction peak can be tuned by coating the nanoparticles with non-conducting shells such as silica, bio molecules, or aluminium oxide.

Effect of Aggregation

When gold nano particles aggregate, the optical properties of the particle change, because the effective particle size, shape, and dielectric environment all change. Gold nanoparticles aggregate particular crown ether

Applications

Electron Microscopy

Colloidal gold and various derivatives have long been among the most widely used labels for antigens in biological electron microscopy. Colloidal gold particles can be attached to many traditional biological probes such as antibodies, lectins, superantigens, glycans, nucleic acids, and receptors. Particles of different sizes are easily distinguishable in electron micrographs, allowing simultaneous multiple-labelling experiments.

In addition to biological probes, gold nanoparticles can be transferred to various mineral substrates, such as mica, single crystal silicon, and atomically flat gold(III), to be observed under atomic force microscopy (AFM).

Medical Research

Drug Delivery System

Gold nanoparticles can be used to optimize the biodistribution of drugs to diseased organs, tissues or cells, in order to improve and target drug delivery. It is important to realize that the nanoparticle-mediated drug delivery is feasible only if the drug distribution is otherwise inadequate. These cases include drug targeting of difficult, unstable molecules (proteins, siRNA, DNA), delivery to the difficult sites (brain, retina, tumors, intracellular organelles) and drugs with serious side effects (e.g. anti-cancer agents). The performance of the nanoparticles depends on the size and surface functionalities in the particles. Also, the drug release and particle disintegration can vary depending on the system (e.g. biodegradable polymers sensitive to pH). An optimal nanodrug delivery system ensures that the active drug is available at the site of action for the correct time and duration, and their concentration should be above the minimal effective concentration (MEC) and below the minimal toxic concentration (MTC).

Gold nanoparticles are being investigated as carriers for drugs such as Paclitaxel. The administration of hydrophobic drugs require molecular encapsulation and it is found that nanosized particles are particularly efficient in evading the reticuloendothelial system.

Gold nanoparticles are also used to circumvent multidrug resistance (MDR) mechanisms. Mechanisms of MDR include decreased uptake of drugs, reduced intracellular drug concentration by activation of the efflux transporters, modifications in cellular pathways by altering cell cycle checkpoints, increased metabolism of drugs, induced emergency response genes to impair apoptotic pathways and altered DNA repair mechanisms.

Tumor Detection

In cancer research, colloidal gold can be used to target tumors and provide detection using SERS (surface enhanced Raman spectroscopy) *in vivo*. These gold nanoparticles are surrounded with Raman reporters, which provide light emission that is over 200 times brighter than quantum dots. It was found that the Raman reporters were stabilized when the nanoparticles were encapsulated with a thiol-modified polyethylene glycol coat. This allows for compatibility and circulation *in vivo*. To specifically target tumor cells, the polyethylenegylated gold particles are conjugated with an antibody (or an antibody fragment such as scFv), against, e.g. epidermal growth factor receptor, which is sometimes overexpressed in cells of certain cancer types. Using SERS, these pegylated gold nanoparticles can then detect the location of the tumor.

Gold nanoparticles accumulate in tumors, due to the leakiness of tumor vasculature, and can be used as contrast agents for enhanced imaging in a time-resolved optical tomography system using short-pulse lasers for skin cancer detection in mouse model. It is found that intravenously administrated spherical gold nanoparticles broadened the temporal profile of reflected optical signals and enhanced the contrast between surrounding normal tissue and tumors.

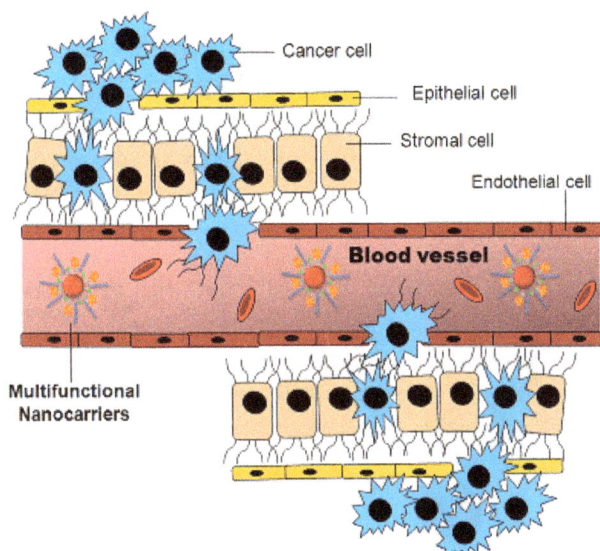

Tumor targeting via multifunctional nanocarriers. Cancer cells reduce adhesion to neighboring cells and migrate into the vasculature-rich stroma. Once at the vasculature, cells can freely enter the bloodstream. Once the tumor is directly connected to the main blood circulation system, multifunctional nanocarriers can interact directly with cancer cells and effectively target tumors.

Therefore, gold nanoparticles have the potential to join numerous therapeutic functions into a single platform, by targeting specific tumor cells, tissues and organs. Actually, Conde et al. reported the evaluation of the inflammatory response and therapeutic siRNA silencing via RGD-nanoparticles in a lung cancer mouse model. This study reported the use of siRNA/RGD gold nanoparticles capable of targeting tumor cells in two lung cancer xenograft mouse models, resulting in successful and significant *c-Myc* oncogene downregulation followed by tumor growth inhibition and prolonged survival of the animals. This delivery system can achieve translocation of siRNA duplexes directly into the tumour cell cytoplasm and accomplish successful silencing of an oncogene expression. Actually, RGD/siRNA-AuNPs can target preferentially and be taken up by tumor cells via integrin αvβ3-receptor-mediated endocytosis with no cytotoxicity, showing that can accumulate in tumor tissues overexpressing αvβ3 integrins and selectively delivered *c-Myc* siRNA to suppress tumor growth and angiogenesis.

Gene Therapy

Gene therapy is receiving increasing attention and, in particular, small-interference RNA (siRNA) shows importance in novel molecular approaches in the knockdown of specific gene expression in cancerous cells. The major obstacle to clinical application is the uncertainty about how to deliver therapeutic siRNAs with maximal therapeutic impact. Gold nanoparticles have shown potential as intracellular delivery vehicles for siRNA oligonucleotides with maximal therapeutic impact.

Recently, Conde et al. provided evidence of *in vitro* and *in vivo* RNAi triggering via the synthesis of a library of novel multifunctional gold nanoparticles, using a hierarchical approach including three biological systems of increasing complexity: *in vitro* cultured human cells, *in vivo* freshwater polyp (*Hydra vulgaris*), and *in vivo* mice models. The authors developed effective conjugation strategies to combine, in a highly controlled way, specific biomolecules to the surface of gold nanoparticles such as: (a) biofunctional spacers: Poly(ethylene glycol) (PEG) spacers used to increase solubility and biocompatibility; (b) cell penetrating peptides such as TAT and RGD peptides: A novel class of membrane translocating agents named cell penetrating peptides (CPPs) that exploit more than one mechanism of endocytosis to overcome the lipophilic barrier of the cellular membranes and deliver large molecules and even small particles inside the cell for their biological actions; and (c) siRNA complementary to a master regulator gene, the protoon-cogene c-myc, were bond covalently (thiol-siRNA) and ionically (naked/unmodified siRNA) to gold nanoparticles.

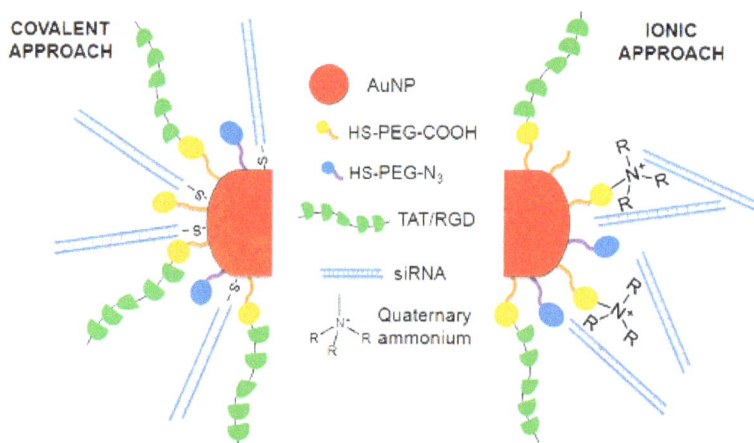

Multifunctional siRNA-gold nanoparticles with several biomolecules: PEG, cell penetration and cell adhesion peptides and siRNA. Two different approaches were employed to conjugate the siRNA to the gold nanoparticle: (1) Covalent approach: use of thiolated siRNA for gold-thiol binding to the nanoparticle; (2) Ionic approach: interaction of the negatively charged siRNA to the modified surface of the AuNP through ionic interactions.

Gold nanoparticles have also shown potential as intracellular delivery vehicles for antisense oligonucleotides (ssDNA,dsDNA) by providing protection against intracellular nucleases and ease of functionalization for selective targeting. Recently, Conde et al. developed a new theranostic system capable of intersecting all RNA pathways: from gene specific downregulation to silencing the silencers, i.e. siRNA and miRNA pathways. The authors reported the development gold nanoparticles functionalized with a fluorophore labeled hairpin-DNA, i.e. gold nanobeacons, capable of efficiently silencing single gene expression, exogenous siRNA and endogenous miRNAs while yielding a quantifiable fluorescence signal directly proportional to the level of silencing. This method describes a gold nanoparticle-based nanobeacon as an innovative theranostic approach for detection and inhibition of sequence-specific DNA and RNA for *in vitro* and *ex vivo* applications. Under hairpin configuration, proximity to gold nanoparticles leads to fluorescence quenching; hybridization to a complementary target restores fluorescence emission due to the gold nanobeacons' conformational reorganization that causes the fluorophore and the gold nanoparticle to part from each other. This concept can easily be extended and adapted to assist the in vitro evaluation of silencing potential of a given sequence to be later used for *ex vivo* gene silencing and RNAi approaches, with the ability to monitor real-time gene delivery action.

Photothermal Agents

Gold nanorods are being investigated as photothermal agents for in-vivo applications. Gold nanorods are rod-shaped gold nanoparticles whose aspect ratios tune the surface plasmon resonance (SPR) band from the visible to near-infrared wavelength. The total extinction of light at the SPR is made up of both absorption and scattering. For the smaller axial diameter nanorods (~10 nm), absorption dominates, whereas for the larger axial diameter nanorods (>35 nm) scattering can dominate. As a consequence, for in-vivo applications, small diameter gold nanorods are being used as photothermal converters of near-infrared light due to their high absorption cross-sections. Since near-infrared light transmits readily through human skin and tissue, these nanorods can be used as ablation components for cancer, and other targets. When coated with polymers, gold nanorods have been observed to circulate in-vivo with half-lives longer than 6 hours, bodily residence times around 72 hours, and little to no uptake in any internal organs except the liver. Apart from rodlike gold nanoparticles, also spherical colloidal gold nanoparticles are recently used as markers in combination with photothermal single particle microscopy.

Radiotherapy dose Enhancer

Following work by Hainfield et al. there has been considerable interest in the use of gold and other heavy-atom containing nanoparticles to enhance the dose delivered to tumors. Since the gold nanoparticles are taken up by the tumors more than the nearby healthy tissue, the dose is selectively enhanced. The biological effectiveness of this type of therapy seems to be due to the local deposition of the radiation dose near the nanoparticles. This mechanism is the same as occurs in heavy ion therapy.

Detection of Toxic Gas

Researchers have developed simple inexpensive methods for on-site detection of hydrogen sulfide H_2S present in air based on the antiaggregation of gold nanoparticles (AuNPs). Dissolving H_2S into a weak alkaline buff solution leads to the formation of HS-, which can stabilize AuNPs and ensure they maintain their red color allowing for visual detection of toxic levels of H_2S.

Gold Nanoparticle based Biosensor

Gold nanoparticles are incorporated into biosensors to enhance its stability, sensitivity, and selectivity. Nanoparticle properties such as small size, high surface-to-volume ratio, and high surface energy allow immobilization of large range of biomolecules. Gold nanoparticle, in particular, could also act as "electron wire" to transport electrons and its amplification effect on electromagnetic light allows it to function as signal amplifiers. Main types of gold nanoparticle based biosensors are optical and electrochemical biosensor.

Optical Biosensor

Gold nanoparticles improve the sensitivity of optical sensor by response to the change in local refractive index. The angle of the incidence light for surface plasmon resonance, an interaction between light wave and conducting electrons in metal, changes when other substances are bounded to the metal surface. Because gold is very sensitive to its surroundings' dielectric constant, binding of an analyte would significantly shift gold nanoparticle's SPR and therefore allow more sensitive detection. Gold nanoparticle could also amplify the SPR signal. When the plasmon wave pass through the gold nanoparticle, the charge density in the wave and the electron I the gold interacted and resulted in higher energy response, so called electron coupling. Since the analyte and bio-receptor now bind to the gold, it increases the apparent mass of the analyte and therefore amplified the signal. These properties had been used to build DNA sensor with 1000-fold sensitive than without the Au NP. Humidity senor was also built by altering the atom interspacing between molecules with humidity change, the interspacing change would also result in a change of the Au NP's LSPR.

Electrochemical Biosensor

Electrochemical sensor covert biological information into electrical signals that could be detected. The conductivity and biocompatibility of Au NP allow it to act as "electron wire". It transfers electron between the electrode and the active site of the enzyme. It could be accomplished in two ways: attach the Au NP to either the enzyme or the electrode. GNP-glucose oxidase monolayer electrode was constructed use these two methods. The Au NP allowed more freedom in the enzyme's orientation and therefore more sensitive and stable detection. Au NP also acts as immobilization platform for the enzyme. Most biomolecules denatures or lose its activity when interacted with the electrode. The biocompatibility and high surface energy of Au allow it to bind to a large amount of protein without altering its activity and results in a more sensitive sensor. Moreover, Au NP also catalyzes biological reactions. Gold nanoparticle under 2 nm has shown catalytic activity to the oxidation of styrene.

Surface Chemistry

In many different types of colloidal gold syntheses, the interface of the nanoparticles can display widely different character – ranging from an interface similar to a self-assembled monolayer to a disordered boundary with no repeating patterns. Beyond the Au-Ligand interface, conjugation of the interfacial ligands with various functional moieties (from small organic molecules to polymers to DNA to RNA) afford colloidal gold much of its vast functionality.

Ligand Exchange/Functionalization

After initial nanoparticle synthesis, colloidal gold ligands are often exchanged with new ligands designed for specific applications. For example, Au NPs produced via the Turkevich-style (or Citrate Reduction) method are readily reacted via ligand exchange reactions, due to the relatively weak binding between the carboxyl groups and the surfaces of the NPs. This ligand exchange can produce conjugation with a number of biomolecules from DNA to RNA to proteins to polymers (such as PEG) to increase biocompatibility and functionality. For example, ligands have been shown to enhance catalytic activity by mediating interactions between adsorbates and the active gold surfaces for specific oxygenation reactions. Ligand exchange can also be used to promote phase transfer of the colloidal particles. Ligand exchange is also possible with alkane thiol-arrested NPs produced from the Brust-type synthesis method, although higher temperatures are needed to promote the rate of the ligand detachment. An alternative method for further functionalization is achieved through the conjugation of the ligands with other molecules, though this method can cause the colloidal stability of the Au NPs to breakdown.

Ligand Removal

In many cases, as in various high-temperature catalytic applications of Au, the removal of the capping ligands produces more desirable physicochemical properties. The removal of ligands from colloidal gold while maintaining a relatively constant number of Au atoms per Au NP can be difficult due to the tendency for these bare clusters to aggregate. The removal of ligands is partially achievable by simply washing away all excess capping ligands, though this method is ineffective in removing all capping ligand. More often ligand removal achieved under high temperature or light ablation followed by washing. Alternatively, the ligands can be electrochemically etched off.

Surface Structure and Chemical Environment

The precise structure of the ligands on the surface of colloidal gold NPs impact the properties of the colloidal gold particles. Binding conformations and surface packing of the capping ligands at the surface of the colloidal gold NPs tend to differ greatly from bulk surface model adsorption, largely due to the high curvature observed at the nanoparticle surfaces. Thiolate-gold interfaces at the nanoscale have been well-studied and the thiolate ligands are observed to pull Au atoms off of the surface of the particles to for "staple" motifs that have significant Thiyl-Au(0) character. The citrate-gold surface, on the other hand, is relatively less-studied due to the vast number of binding conformations of the citrate to the curved gold surfaces. A study performed in 2014 identified that the most-preferred binding of the citrate involves two carboxylic acids and the hydroxyl group of the citrate binds three surface metal atoms.

Toxicity

As gold nanoparticles (AuNPs) are further investigated for targeted drug delivery in humans, their toxicity needs to be considered. For the most part, it is suggested that AuNPs are biocompatible, but it is important to ask at what concentration they would be toxic, and if that concentration falls within the range of used concentrations. Toxicity can be tested *in vitro* and *in vivo*. *In vitro* toxicity results can vary depending on the type of the cellular growth media with different protein compositions, the method used to determine cellular toxicity (cell health, cell stress, how many cells are taken into a cell), and the capping ligands in solution. *In vivo* assessments can determine the general health of an organism (abnormal behavior, weight loss, average life span) as well as tissue specific toxicology (kidney, liver, blood) and inflammation and oxidative responses. *In vitro* experiments are more popular than *in vivo* experiments because *in vitro* experiments are more simplistic to perform than *in vivo* experiments.

Toxicity and Hazards in Synthesis

While AuNPs themselves appear to have low or negligible toxicity, and the literature shows that the toxicity has much more to do with the ligands rather than the particles themselves, the synthesis of them involves chemicals that are hazardous. Sodium borohydride, a harsh reagent, is used to reduce the gold ions to gold metal. The gold ions usually come from chloroauric acid, a potent acid. Because of the high toxicity and hazard of reagents used to synthesize AuNPs, the need for more "green" methods of synthesis arose.

Toxicity due to Capping Ligands

Some of the capping ligands associated with AuNPs can be toxic while others are nontoxic. In gold nanorods (AuNRs), it has been shown that a strong cytotoxicity was associated with CTAB-stabilized AuNRs at low concentration, but it is thought that free CTAB was the culprit in toxicity . Modifications that overcoat these AuNRs reduces this toxicity in human colon cancer cells (HT-29) by preventing CTAB molecules from desorbing from the AuNRs back into the solution. Ligand toxicity can also be seen in AuNPs. Compared to the 90% toxicity of HAuCl4 at the same concentration, AuNPs with carboxylate termini were shown to be non-toxic. Large AuNPs conjugated with biotin, cysteine, citrate, and glucose were not toxic in human leukemia cells (K562) for concentrations up to 0.25 M. Also, citrate-capped gold nanospheres (AuNSs) have been proven to be compatible with human blood and did not cause platelet aggregation or an immune response. However, citrate-capped gold nanoparticles sizes 8-37 nm were found to be lethally toxic for mice, causing shorter lifespans, severe sickness, loss of appetite and weight, hair discoloration, and damage to the liver, spleen, and lungs; gold nanoparticles accumulated in the spleen and liver after traveling a section of the immune system. There are mixed-views for polyethylene glycol (PEG)-modified AuNPs. These AuNPs were found to be toxic in mouse liver by injection, causing cell death and minor inflammation. However, AuNPs conjugated with PEG copolymers showed negligible toxicity towards human colon cells (Caco-2). AuNP toxicity also depends on the overall charge of the ligands. In certain doses, AuNSs that have positively-charged ligands are toxic in monkey kidney cells (Cos-1), human red blood cells, and E. coli because of the AuNSs interaction with the negatively-charged cell membrane; AuNSs with negatively-charged ligands have been found to be nontoxic in these species. In addition to the previously mentioned "in vivo" and "in vitro" experiments, other similar experiments have been performed. Alkylthiolate-AuNPs with trimethylammonium ligand termini mediate the translocation of DNA across mammalian cell membranes "in vitro" at a high level, which is detrimental to these cells. Corneal haze in rabbits have been healed "in vivo" by using polyethylemnimine-capped gold nanoparticles that were transfected with a gene that promotes wound healing and inhibits corneal fibrosis.

Toxicity due to Size of Nanoparticles

Toxicity in certain systems can also be dependent on the size of the nanoparticle. AuNSs size 1.4 nm were found to be toxic in human skin cancer cells (SK-Mel-28), human cervical cancer cells (HeLa), mouse fibroblast cells (L929), and mouse macrophages (J774A.1), while 0.8, 1.2, and 1.8 nm sized AuNSs were less toxic by a six-fold amount and 15 nm AuNSs were nontoxic. There is some evidence for AuNP buildup after injection in "in vivo" studies, but this is very size dependent. 1.8 nm AuNPs were found to be almost totally trapped in the lungs of rats. Different sized AuNPs were found to buildup in the blood, brain, stomach, pancreas, kidneys, liver, and spleen.

Synthesis

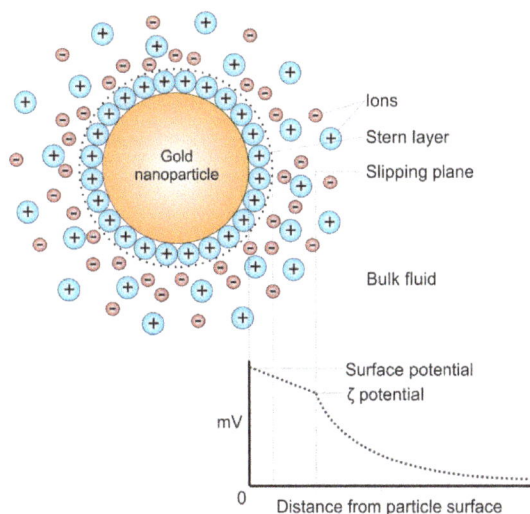

Potential difference as a function of distance from particle surface.

Generally, gold nanoparticles are produced in a liquid ("liquid chemical methods") by reduction of chloroauric acid ($H[AuCl_4]$). After dissolving $H[AuCl_4]$, the solution is rapidly stirred while a reducing agent is added. This causes Au^{3+} ions to be reduced to Au^+ions. Then a disproportionation reaction occurs whereby 3 Au^+ ions give rise to Au^{3+} and 2 Au^0 atoms. The Au^0 atoms act as center of nucleation around which further Au^+ ions gets reduced. To prevent the particles from aggregating, some sort of stabilizing agent that sticks to the nanoparticle surface is usually added. In the Turkevich method of Au NP synthesis, citrate initially acts as the reducing agent and finally as the capping agent which stabilizes the Au NP through electrostatic interactions between the lone pair of electrons on the oxygen and the metal surface. Also, gold colloids can be synthesised without stabilizers by laser ablation in liquids.

They can be functionalized with various organic ligands to create organic-inorganic hybrids with advanced functionality.

Turkevich Method

The method pioneered by J. Turkevich et al. in 1951 and refined by G. Frens in the 1970s, is the simplest one available. In general, it is used to produce modestly monodisperse spherical gold nanoparticles suspended in water of around 10–20 nm in diameter. Larger particles can be produced, but this comes at the cost of monodispersity and shape. It involves the reaction of small amounts of hot chloroauric acid with small amounts of sodium citrate solution. The colloidal gold will form because the citrate ions act as both a reducing agent and a capping agent. A capping agent is used in nanoparticle synthesis to stop particle growth and aggregation. A good capping agent has a high affinity for the new nuclei so it will bind to surface atoms which stabilizes the surface energy of the new nuclei and makes so that they cannot bind to other nuclei.

Recently, the evolution of the spherical gold nanoparticles in the Turkevich reaction has been elucidated. It is interesting to note that extensive networks of gold nanowires are formed as a transient intermediate. These gold nanowires are responsible for the dark appearance of the reaction solution before it turns ruby-red.

To produce larger particles, less sodium citrate should be added (possibly down to 0.05%, after which there simply would not be enough to reduce all the gold). The reduction in the amount of sodium citrate will reduce the amount of the citrate ions available for stabilizing the particles, and this will cause the small particles to aggregate into bigger ones (until the total surface area of all particles becomes small enough to be covered by the existing citrate ions).

Brust-Schiffrin Method

This method was discovered by Brust and Schiffrin in the early 1990s, and can be used to produce gold nanoparticles in organic liquids that are normally not miscible with water (like toluene). It involves the reaction of a chlorauric acid solution with tetraoctylammonium bromide (TOAB) solution in toluene and sodium borohydride as an anti-coagulant and a reducing agent, respectively.

Here, the gold nanoparticles will be around 5–6 nm. $NaBH_4$ is the reducing agent, and TOAB is both the phase transfer catalyst and the stabilizing agent.

It is important to note that TOAB does not bind to the gold nanoparticles particularly strongly, so the solution will aggregate gradually over the course of approximately two weeks. To prevent this, one can add a stronger binding agent, like a thiol (in particular, alkanethiols), which will bind to gold, producing a near-permanent solution. Alkanethiol protected gold nanoparticles can be precipitated and then redissolved. Thiols are better binding agents because there is a strong affinity for the gold-sulfur bonds that form when the two substances react with each other. Tetra-dodecanthiol is a commonly used strong binding agent to synthesize smaller particles. Some of the phase transfer agent may remain bound to the purified nanoparticles, this may affect physical properties such as solubility. In order to remove as much of this agent as possible, the nanoparticles must be further purified by soxhlet extraction.

Perrault Method

This approach, discovered by Perrault and Chan in 2009, uses hydroquinone to reduce $HAuCl_4$ in an aqueous solution that contains 15 nm gold nanoparticle seeds. This seed-based method of synthesis is similar to that used in photographic film development, in which silver grains within the film grow through addition of reduced silver onto their surface. Likewise, gold nanoparticles can act in conjunction with hydroquinone to catalyze reduction of ionic gold onto their surface. The presence of a stabilizer such as citrate results in controlled deposition of gold atoms onto the particles, and growth. Typically, the nanoparticle seeds are produced using the citrate method. The hydroquinone method complements that of Frens, as it extends the range of monodispersed spherical particle sizes that can be produced. Whereas the Frens method is ideal for particles of 12–20 nm, the hydroquinone method can produce particles of at least 30–300 nm.

Martin Method

This simple method, discovered by Martin and Eah in 2010, generates nearly monodisperse "naked" gold nanoparticles in water. Precisely controlling the reduction stoichiometry by adjusting the ratio of $NaBH_4$-NaOH ions to $HAuCl_4$-HCl ions within the "sweet zone," along with heating, enables reproducible diameter tuning between 3–6 nm. The aqueous particles are colloidally stable due to their high charge from the excess ions in solution. These particles can be coated with various hydrophilic functionalities, or mixed with hydrophobic molecules for applications in non-polar solvents. In non-polar solvents the nanoparticles remain highly charged, and self-assemble on liquid droplets to form 2D monolayer films of monodisperse nanoparticles.

Nanotech Applications

Bacillus licheniformis can be used in synthesis of gold nanocubes with sizes between 10 and 100 nanometres. Gold nanoparticles are usually synthesized at high temperatures in organic solvents or using toxic reagents. The bacteria produce them in much milder conditions.

Navarro et al. Method

The precise control of particle size with a low polydispersity of spherical gold nanoparticles remains difficult for particles larger than 30 nm. In order to provide maximum control on the NP structure, Navarro and co-workers used a modified Turkevitch-Frens procedure using sodium acetylacetonate (Na(acac)) as the reducing agent and sodium citrate as the stabilizer.

Sonolysis

Another method for the experimental generation of gold particles is by sonolysis. The first method of this type was invented by Baigent and Müller. This work pioneered the use of ultrasound to provide the energy for the processes involved and allowed the creation of gold particles with a diameter of under 10 nm. In another method using ultrasound, the reaction of an aqueous solution of $HAuCl_4$ with glucose, the reducing agents are hydroxyl radicals and sugar pyrolysis radicals (forming at the interfacial region between the collapsing cavities and the bulk water) and the morphology obtained is that of nanoribbons with width 30–50 nm and length of several micrometers. These ribbons are very flexible and can bend with angles larger than 90°. When glucose is replaced by cyclodextrin (a glucose oligomer), only spherical gold particles are obtained, suggesting that glucose is essential in directing the morphology toward a ribbon.

Block Copolymer-mediated Method

An economical, environmentally benign and fast synthesis methodology for gold nanoparticles using block copolymer has been developed by Sakai et al. In this synthesis methodology, block copolymer plays the dual role of a reducing agent as well as a stabilizing agent. The formation of gold nanoparticles comprises three main steps: reduction of gold salt ion by block copolymers in the solution and formation of gold clusters, adsorption of block copolymers on gold clusters and further reduction of gold salt ions on the surfaces of these gold clusters for the growth of gold particles in steps, and finally its stabilization by block copolymers. But this method usually has a limited-yield (nanoparticle concentration), which does not increase with the increase in the gold salt concentration. Recently, Ray et al. demonstrated that the presence of an additional reductant (trisodium citrate) in 1:1 mole ratio with gold salt enhances the yield by manyfold at ambient conditions and room temperature.

"Green Chemistry" Based Methods

Methods Employing Phytochemicals

Phytochemicals found in various plant sources have been utilized as a means of developing a more economical and environmentally friendly synthetic pathway in the formation of gold nanoparticles. In accordance to the principles of "green chemistry," these methods employ the use of non-toxic chemicals, marginal energy consumption, renewable materials, and environmentally benign solvents to minimize the use, disposal, and health repercussions of hazardous chemicals. Additionally, these methods provide a more efficient synthetic pathway through a one-step process without the use of supplementary surfactants or polymers, capping agents, or templates to restrict agglomeration of the gold nanoparticles. This method is effective in producing well-defined gold nanoparticles since the phytochemicals perform a dual role as both a reducing agent of gold and as a stabilizer in the formation of a sturdy coating on the nanoparticles. One "green" method that has been employed in the formation of gold nanoparticles utilizes the phytochemicals and polyphenols in Darjeeling black tea leaves, with water acting as a benign solvent at room temperature. The phytochemicals in the black tea reduce $HAuCl_4$ and stabilize the aggregation of the gold atoms as the nanoparticle is formed. In addition to the "green" benefits of using black tea, the size of the nanoparticle is influenced by the concentration of the tea, and the absorbance and size of the gold nanoparticles formed can be easily determined using UV-Vis spectrometry abiding that an increase in λ_{max} correlates to an increase in the size of the nanoparticle.

Other "green" methods that have been studied and employed include the use of *Elettaria cardamomum* (cardamom) and cinnamon in the synthesis of gold nanoparticles, as well as *Syzygium aromaticum* (cloves) in the formation of copper nanoparticles and table sugar in the formation of silver nanoparticles.

Gold Nanoparticles as a Benign Starting Material to Access Gold Sponges

Besides using phytochemicals from plant sources to act as reducing agents and stabilizing agents, several other approaches have been taken to achieve more "green" approaches to gold nanoparticle syntheses. One such approach employs thiolated poly(ethylene glycol) (PEG Thiol) to destabilize gold nanoparticles prepared by citrate reduction so that they self-assemble into mesoporous gold sponges. Mesoporous gold sponges are attractive materials for molecular sensing by Surface-Enhanced Raman Spectroscopy (SERS), for catalysis, and for fuel cell construction. The following approach is "green" because PEG Thiol is biocompatible, and because it requires relatively little energy; PEG Thiol-triggered self-assembly of mesoporous gold sponges occurs at room temperature. By contrast, the most popular method of generating mesoporous gold sponges, dealloying Au-Ag alloys, employs electrochemical corrosion.

Gold Nanoparticle Synthesis in Flow

Another green approach is a modification to the Turkevich citrate reduction making use of flow chemistry, reported by Bayazit et al. Flow chemistry is an appealing replacement for many heated batch reactions. By exposing more surface area of the reaction to the heating element, flow reactors heat a reaction faster and more evenly than a batch reactor can, promoting rapid nucleation and smaller particle sizes with higher monodispersity.

Characterization

The future of nanotechnology rests upon approaches to making new, useful nanomaterials and testing them in complex systems. However, precise and trusting nanoparticle formation can only be promised after their thorough characterization. Here some of the basic physical and biophysical techniques are briefed which will be central to nanoparticles research.

X-ray Diffraction

The genesis of XRD can be traced to the suggestion of Max von Laue in 1912 that a crystal can be considered as a three-dimensional diffraction grating. It is generally used for qualitative analysis.X-ray diffraction (XRD) method is the most basic method for characterizing the crystal structures. X-rays corresponds to electromagnetic radiation in the wavelength range of 1 Å. The wavelength range is below that of ultraviolet light and above that of gamma rays.

X-rays are generally produced when electrons of several thousands of electron volts are decelerated or stopped by metals. This will produce a white radiation up to a threshold frequency corresponding to the kinetic energy of the particle. This threshold corresponds to a wavelength (in angstroms)

$$\lambda = 12399/V$$

Where, V- accelerating voltage of the electrons.

When electrons fall on matter with high energy, electrons can be ejected from various energy levels. Electron ejection from the core orbital is also accompanied by the emission of characteristic X-rays.

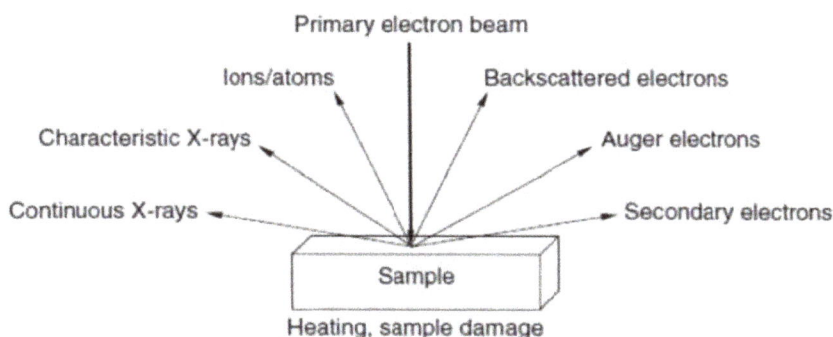

Electron beam induced processes in the sample

XRD method is based on the measurements of X-ray intensities scattered by the statistically distributed electrons belonging to the atoms in the material. Since the most stable structure of a pure material is crystal, where the atoms are periodically arranged, the pattern of the positions and intensities of XRD peaks can be uniquely assigned to the material. Therefore, XRD measurement is important to identify the main component of materials. The diffraction angle 2θ, the interplanar distance d and X-ray wavelength λ are connected with each other by the Bragg's law:

$$n\lambda = 2d\sin\theta$$

The X-ray diffraction experiment requires the following: a radiation, a sample and a detector for the reflected radiation. In each of these cases, there can be several variations.

In the Debye-Scherrer method of diffraction, we use a monochromatic X-ray and a powder sample with every possible set of lattice planes exposed to the radiation.

In the modern diffraction method called diffractometry, a convergent beam strikes the sample and the intensity as a function of diffraction angle is measured. The position of the diffraction peak and the intensity at this point are the two factors used in the determination. Both these can be measured accurately and compared with standards in the literature.

The particle size is obtained as broadening of the diffracted lines and is given by the Scherrer formula,

$$t = 0.9\lambda / (B\cos\theta)$$

Where t -thickness of the crystallite in (angstroms)

θ-Bragg's angle.

B –peak broadening; full width at intensity half maxima

The width of the diffraction curve (B) increases as the thickness of the crystal decreases.

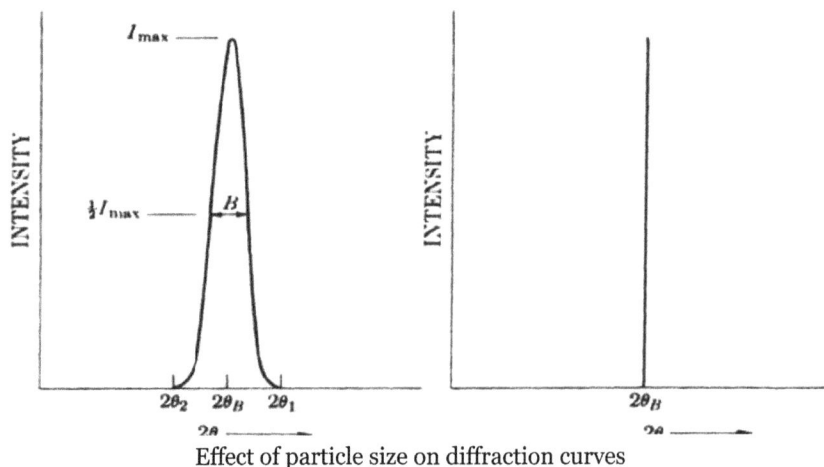

Effect of particle size on diffraction curves

Thus, information obtained from XRD can be used to determine the crystal structure of the sample.

Microscopy

Electron Microscopy

Transmission electron microscopy

The properties of polycrystalline material are different from that of single crystalline material, due to the grain boundaries and their non-periodic arrangements of atoms TEM plays important roles for characterization of grain boundaries and assists the development of new polycrystalline materials. TEM is an equipment to let the incident electron beam to transmit a thin specimen at high-acceleration voltage (80–3,000 kV) which results in generating signals caused by the interaction between the specimen and incident electrons. Structures, compositions and chemical bonds of the specimen can be determined from these signals. In general, there are three types of transmitted electrons observed by TEM.

- Unscattered electrons- are caused by incident electrons transmitted through the thin specimen without any interaction occurring inside the specimen. Amount of unscattered electrons is inversely proportional to the specimen thickness.

- Elastically scattered electrons- are caused by the incident electrons that are scattered by atoms in the specimen without losing energy. These elastically scattered electrons follow the Bragg's diffraction law. All incident electrons have the same energy and enter the specimen normal to its surface. All electrons that are scattered by the same atomic spacing will be scattered by the same angle. These "the same angle" scattered electrons are gathered by lens and form a pattern of spots; each spot corresponding to a specific atomic spacing. This diffracted pattern yields information about the orientation, atomic arrangements and phases present in the region of interest.

- Inelastically scattered electrons- are caused by the incident electrons that interact with atoms in specimen with losing their energy. These provide two types of information.

   a. The inelastic loss of energy by the incident electrons, characteristic of the elements. These energies are unique to each bonding state of each element and thus can be used to extract both compositional and chemical bonding information of the specimen.

   b. Another one is the formation of bands with alternating light and dark lines, known as Kikuchi bands. These bands are also formed by inelastic scattering interactions related to the atomic spacing in the specimen

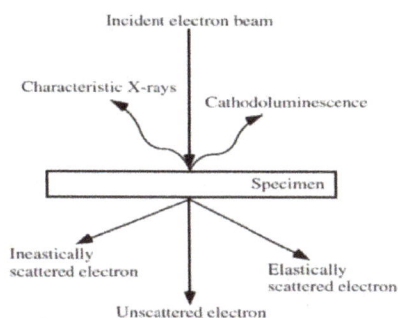

Interaction between incident electron beam and specimen in case of TEM

There are two main mechanisms of contrast in an image. The transmitted and scattered beams can be recombined at the image plane, thus preserving their amplitudes and phases. This results in the phase contrast image of the object. An amplitude contrast image can be obtained by eliminating the diffracted beams. This is achieved by placing suitable apertures below the back focal plane of the objective lens. This image is called the bright field image. The size of the objective aperture should be small enough to remove all diffracted electron beams caused by the specimen One can also exclude all other beams except the particular diffracted beam of interest. The image using this is called the dark field image. The advantage of the dark-field imaging method is its high-diffraction contrast. The dark-field imaging technique is usually used for observing grain size distributions and dislocations.

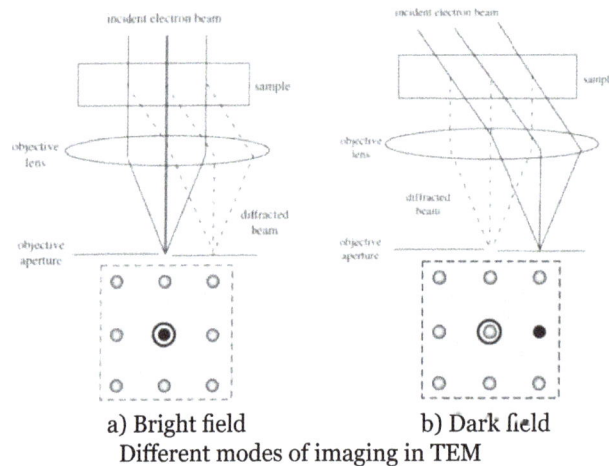

a) Bright field   b) Dark field
Different modes of imaging in TEM

Scanning Electron Microscopy

Scanning electron microscope (SEM) uses a focused beam of high-energy electrons to generate a variety of signals at the surface of solid specimens. The signals that derive from electron-sample interactions reveal information about the sample including external morphology (texture), chemical composition, and crystalline structure and orientation of materials making up the sample. In most applications, data are collected over a selected area of the surface of the sample, and a 2-dimensional image is generated that displays spatial variations in these properties. The SEM is also capable of performing analyses of selected point locations on the sample; this approach is especially useful in qualitatively or semi-quantitatively determining chemical compositions (using Energy Dispersive analysis).

The interactions exploited for SEM analysis include-

- Secondary electrons (that produce SEM images)- most valuable for showing morphology and topography on samples

- Backscattered electrons - illustrating contrasts in composition in multiphase samples

- Diffracted backscattered electrons- are used to determine crystal structures and orientations of minerals

SEM analysis is considered to be "non-destructive"; that is, x-rays generated by electron interactions do not lead to volume loss of the sample, so it is possible to analyze the same materials repeatedly.

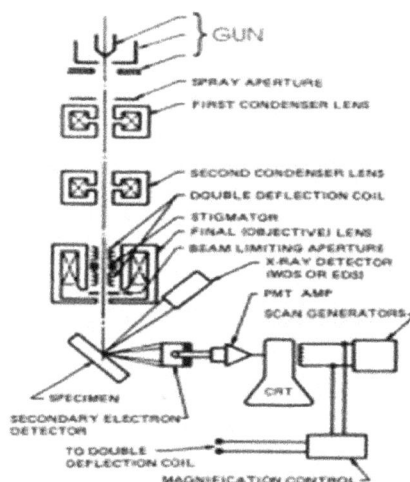

Schematic representation of SEM

Essential components of all SEMs include the following-

- Electron Source ("Gun")
- Electron Lenses
- Sample Stage
- Detectors for all signals of interest
- Display / Data output devices
- Infrastructure Requirements:
  o Power Supply
  o Vacuum System
  o Cooling system
  o Vibration-free floor
  o Room free of ambient magnetic and electric fields

Sample preparation for SEM is a crucial task in particular for insulating samples. Most electrically insulating samples are coated with a thin layer of conducting material, commonly carbon, gold, or some other metal or alloy. The choice of material for conductive coatings depends on the data to be acquired for eg.-carbon is most desirable if elemental analysis is a priority, while metal coatings are most effective for high resolution electron imaging applications.

SEM finds use for varied applications some of them are-

1. Routinely used to generate high-resolution images of shapes of objects

2. To show spatial variations in chemical compositions

3. To identify phases based on qualitative chemical analysis and/or crystalline structure.

4. Precise measurement of very small features and objects down to 50 nm in size is also accomplished using the SEM.

SEM offers advantage of wide applicability and rapid data acquisition but suffers from drawbacks of high vacuum requirement of the order 10^{-5} - 10^{-6} torr.

Atomic Force Microscopy

AFM provides a 3D profile of the surface on a nanoscale, by measuring forces between a sharp probe (<10 nm) and surface at very short distance (0.2-10 nm, probe-sample separation). The probe is supported on a flexible cantilever. The AFM tip "gently" touches the surface and records the small force between the probe and the surface.

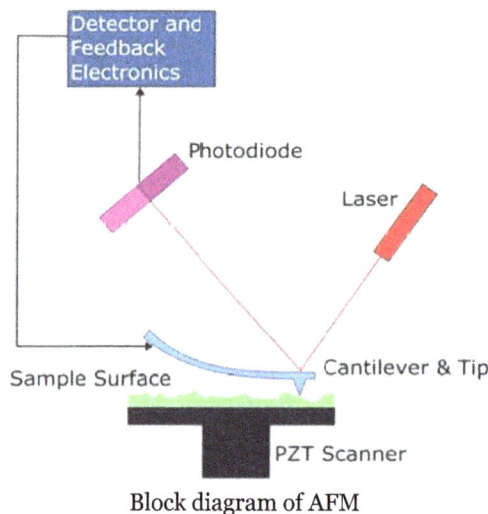

Block diagram of AFM

The probe is placed on the end of a cantilever (which one can think of as a spring). The amount of force between the probe and is dependent on the spring constant (stiffness of the cantilever and the distance between the probe and the sample surface. This force can be described using Hooke's Law:)

$$F = -k \cdot x$$

F = Force

k = spring constant

x = cantilever deflection

If the spring constant of cantilever (typically ~ 0.1-1 N/m) is less than surface, the cantilever bends and deflection is monitored. Probes are typically made from Si_3N_4, or Si. Different cantilever lengths, materials, and shapes allow for varied spring constants and resonant frequencies. The motion of the probe across the surface is controlled using feedback loop and piezoelectronic scanners. There are 3 primary imaging modes in AFM-

(1) Contact AFM (< 0.5 nm probe-surface separation)

When cantilever bends are due to sample interaction the force on the tip is repulsive. By maintaining a constant cantilever deflection (using the feedback loops) the force between the probe and the sample remains constant and an image of the surface is obtained.

Advantages: fast scanning, good for rough samples, used in friction analysis

Disadvantages: at time forces can damage/deform soft samples (however imaging in liquids often resolves this issue)

(2) Intermittent contact (0.5-2 nm probe-surface separation)

However, in this mode the cantilever is oscillated at its resonant frequency, Figure. The probe lightly "taps" on the sample surface during scanning, contacting the surface at the bottom of its swing. By maintaining constant oscillation amplitude a constant tip-sample interaction is maintained and an image of the surface is obtained.

Advantages: allows high resolution of samples that are easily damaged and/or loosely held to a surface; good for biological samples

Disadvantages: more challenging to image in liquids, slower scan speeds needed

(3) Non-contact AFM (0.1-10 nm probe-surface separation)

The probe does not contact the sample surface, but oscillates above the adsorbed fluid layer on the surface during scanning. (Note: all samples unless in a controlled UHV or environmental chamber have some liquid adsorbed on the surface). Using a feedback loop to monitor changes in the amplitude due to attractive VdW forces the surface topography can be measured.

Advantages: very low force exerted on the sample (10-12N), extended probe lifetime

Disadvantages: generally lower resolution; contaminant layer on surface can interfere with oscillation; usually need ultra-high vacuum (UHV) to have best imaging.

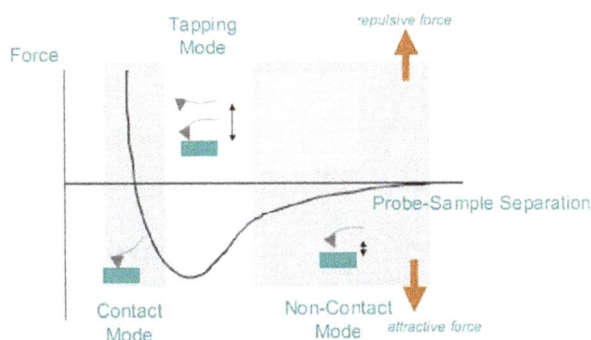

Plot of force as a function of probe-sample separation

Spectroscopy

UV-Visible Spectroscopy

The word 'spectroscopy' is used as a collective term for all the analytical techniques based on the interaction of light and matter. Spectrophotometry is one of the branches of spectroscopy where we measure the absorption of light by molecules that are in a gas or vapour state or dissolved molecules/ions. Spectrophotometry investigates the absorption of the different substances between the wavelength limits 190 nm and 780 nm (visible spectroscopy is restricted to the wavelength range of electromagnetic radiation detectable by the human eye, that is above ~360 nm; ultraviolet spectroscopy is used for shorter wavelengths). In this wavelength range the absorption of the electromagnetic radiation is caused by the excitation (i.e. transition to a higher energy level) of the bonding and non-bonding electrons of the ions or molecules. A graph of absorbance against wavelength gives the sample's absorption spectrum.

Spectrophotometry is used for both qualitative and quantitative investigations of samples. The wavelength at the maximum of the absorption band gives information about the structure of the molecule or ion and the extent of the absorption is proportional with the amount of the species absorbing the light.

Quantitative measurements are based on Beer's Law which is described as follows:

$$A = \varepsilon\, c\, l$$

Where; A-absorbance [$A = \log_{10}(I_0 / I), I_0$ is the incident light's intensity and I is the light intensity after it passes through the sample];

e- Molar absorbance or absorption coefficient [in $dm^3\ mol^{-1}\ cm^{-1}$ units];

c- Concentration (molarity) of the compound in the solution [in $moldm^{-3}$ units];

l - Path length of light in the sample [in cm units].

The instruments used for spectrophotometry are called photometers and Spectrophotometers. The difference between them is that we can only make measurements at a particular wavelength with a photometer, but spectrophotometers can be used for the whole wavelength range. Both types of instruments have suitable light sources, monochromator (that selects the light with the necessary wavelength) and a detector. The solution is put into a sample tube (called a "cuvette"). The light intensity measured by the detector is converted into an electric signal and is displayed as a certain absorbance on the readout.

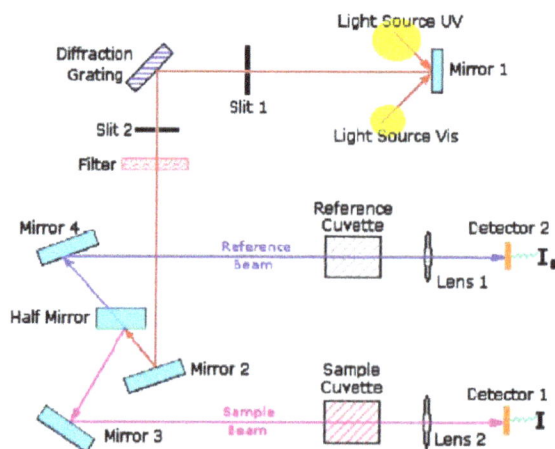

UV/Vis spectrophotometer

Photoluminescence Spectroscopy

Photoluminescence spectroscopy is a contactless, nondestructive method of probing the electronic structure of materials. Light is directed onto a sample, where it is absorbed and imparts excess energy into the material in a process called photo-excitation. One way this excess energy can be dissipated by the sample is through the emission of light, or luminescence. In the case of photo-excitation, this luminescence is called photoluminescence. The intensity and spectral content of this photoluminescence is a direct measure of various important material properties.

Photo-excitation causes electrons within the material to move into permissible excited states. When these electrons return to their equilibrium states, the excess energy is released and may include the emission of light (a radiative process) or may not (a nonradiative process). The energy of the emitted light (photoluminescence) relates to the difference in energy levels between the two electron states involved in the transition between the excited state and the equilibrium state. The quantity of the emitted light is related to the relative contribution of the radiative process. Photoluminescence is used for band gap determination particularly in case of semiconductors.

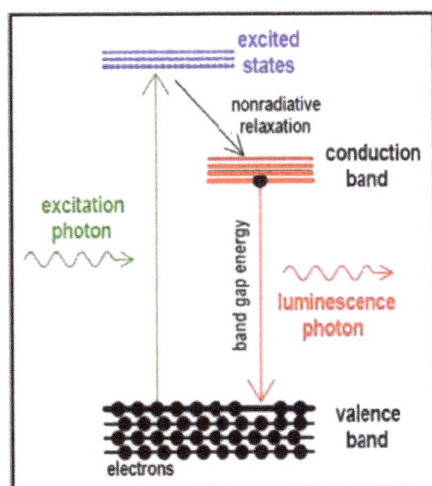

Photoluminescence

The light from an excitation source passes through a filter or monochromator, and strikes the sample. A proportion of the incident light is absorbed by the sample, and some of the molecules in the sample fluoresce. The fluorescent light is emitted in all directions. Some of this fluorescent light passes through a second filter or monochromator and reaches a detector (photosensor), which is usually placed at 90° to the incident light beam to minimize the risk of transmitted or reflected incident light reaching the detector. Various light sources may be used as excitation sources; including lasers, photodiodes, and lamps; xenon arcs and mercury-vapor lamps in particular. Filters and/or monochromators may be used in fluorimeters. A monochromator transmits light of an adjustable wavelength with an adjustable tolerance. The most common type of monochromator utilizes a diffraction grating, that is, collimated light illuminates a grating and exits with a different angle depending on the wavelength. The monochromator can then be adjusted to select which wavelengths to transmit. As mentioned before, the fluorescence is most often measured at a 90° angle relative to the excitation light. This geometry is used instead of placing the sensor at the line of the excitation light at a 180° angle in order to avoid interference of the transmitted excitation light. The detector can either be single-channeled or multichanneled.

References

- Graf, Christina; Vossen, Dirk L.J.; Imhof, Arnout; van Blaaderen, Alfons (July 11, 2003). "A General Method To Coat Colloidal Particles with Silica". Langmuir. 19 (17): 6693–6700. doi:10.1021/la0347859

- Theory, Production and Mechanism of Formation of Monodispersed Hydrosols, Victor K. LaMer and Robert H. Dinegar, Journal of the American Chemical Society 1950 72 (11), 4847-4854

- Electron Microscopy, 2nd Edition, by John J. Bozzola, Jones & Bartlett Publishers; 2 Sub edition (October 1998) ISBN 0-7637-0192-0

- Dong, X.; Ji, X.; Jing, J.; Li, M.; Li, J.; Yang, W. (2010). "Synthesis of Triangular Silver Nanoprisms by Stepwise Reduction of Sodium Borohydride and Trisodium Citrate". J. Phys. Chem. C. 114 (5): 2070–2074. doi:10.1021/jp909964k

- Chuang; et al. "Allergenicity and toxicology of inhaled silver nanoparticles in allergen-provocation mice models". International Journal of Nanomedicine. 2013 (8): 4495–4506. doi:10.2147/IJN.S52239

- Practical Electron Microscopy: A Beginner's Illustrated Guide, by Elaine Evelyn Hunter. Cambridge University Press; 2nd edition (September 24, 1993) ISBN 0-521-38539-3

- Shan, Z.; Wu, J.; Xu, F.; Huang, F.-Q.; Ding, H. (2008). "Highly Effective Silver/Semiconductor Photocatalytic Composites Prepared By a Silver Mirror Reaction". J. Phys. Chem. C. 112 (39): 15423–15428. doi:10.1021/jp804482k

- Hirai; et al. (2014). "Silver nanoparticles induce silver nanoparticle-specific allergic responses (HYP6P.274)". The Journal of Immunology. 192 (118): 19

- Electron Microscopy: Methods and Protocols (Methods in Molecular Biology), by John Kuo (Editor). Humana Press; 2nd edition (February 27, 2007) ISBN 1-58829-573-7

- Pietrobon B., Mceachran M., Kitaev V. (2009). "Synthesis Of Size-Controlled Faceted Pentagonal Silver Nanorods with Tunable Plasmonic Properties and Self-Assembly of These Nanorods". ACS Nano. 3: 21–26. doi:10.1021/nn800591y

- Parkes, A. (2006). "Silver-coated dressing Acticoat". Journal of Trauma-Injury Infection & Critical Care. 61 (1): 239–40. doi:10.1097/01.ta.0000224131.40276.14

- L. He, M.D. Musick, S. R. Nicewarner, F.G. Salinas, (title missing) Journal of the American Chemical Society, 2000, 122, 9071

- Mohan, R.R.; et., al. (June 2013). "BMP7 Gene Transfer via Gold Nanoparticles into Stroma Inhibits Corneal Fibrosis In Vivo". Plos One. 8 (6): 1–9. doi:10.1371/journal.pone.0066434

3

Detection Technology

Nanotechnology has contributed in the development of sensors. It outweighs conventional sensors as it has higher sensitivity, consumes less power, effectively cuts production cost, and provides better stability. Biosensor, a device to detect analytes in biological systems, has gained immense popularity. Tools are an important component of any field of study. The following chapter elucidates the various tools that are related to the nanobiotechnology.

Transducing Element

Nanotechnology helps in development of small, highly-efficient and inexpensive sensors, with broad applications. These offer significant advantages over conventional sensors. This includes greater sensitivity and selectivity, lower production costs, reduced power consumption as well as improved stability.

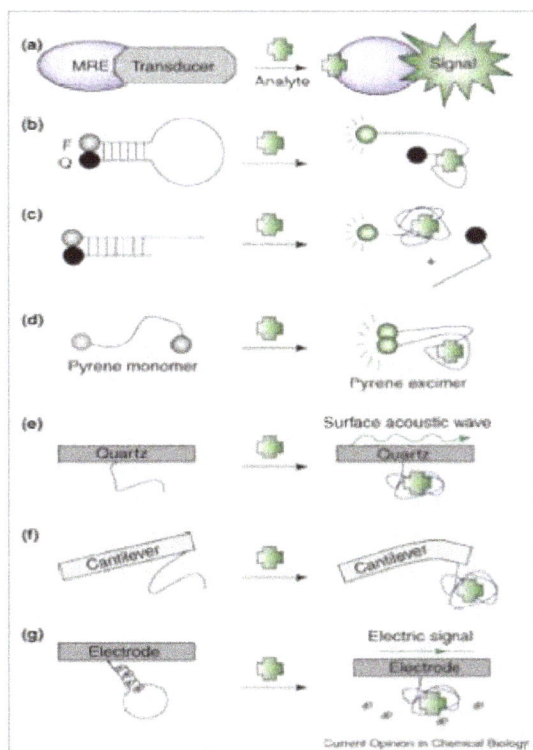

Due to above mentioned characteristics, bio nano-sensors are gaining lot of attraction in diagnosis and other areas of sensing where minute quantities of analyte are undetectable by conventional sensors. DNA detection using biosensors is also gaining attention as non-pcr methods for detection can be implied using bio transducers.

Biosensor: A biosensor is a device used to detect the presence of an analyte in biological systems. It consists of two parts:

1. Molecular recognition element: it is the molecule or compound which reacts with the analyte or binds to the analyte and confirms its presence by sending a signal in the form of light, heat, sound, mechanical deformation etc.

2. Transducer: it is the device which is attached to the molecular recognition element which transforms the signal of any form i.e. heat, light, mechanical strain etc. to a perceivable electrical signal.

In this report the importance of transducers is specifically discussed. Various types of transducing elements, their principle and applications in biotechnology are briefly explained.

Calorimetric Transducer

A calorimetric transducer, as the name suggests, converts heat changes into electrical signals. Therefore, many endothermic or exothermic biochemical reactions can be detected using a calorimetric biosensor. The transducer consists of a thermistor which is a resistor whose resistance varies with the change in temperature and is able to detect very small changes in temperature. This way it serves as a beacon for the presence of an analyte. The reactions are carried out in a controlled and closed reactor to ensure minimum amount of heat loss.

Most of the enzyme catalyzations are exothermic in nature. And hence the heat evolved can be carefully noted by the change in current in the installed circuit. The following picture shows a calorimetric transducer:

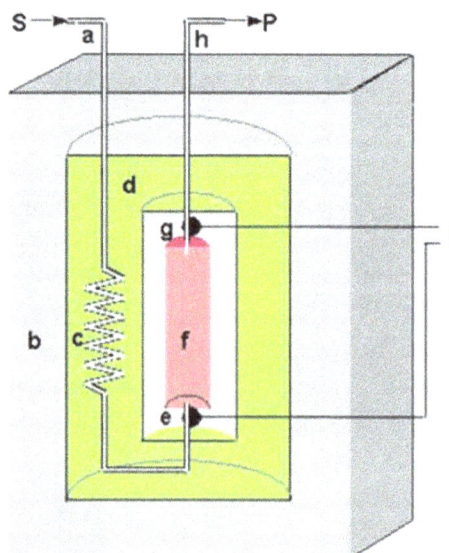

The enzyme is made to enter the highly controlled system through a. Then it passes to the heat exchanger c. Through the first thermistor e, it reaches the reactor where it is made to react with the anlyte. Then, it is removed out from h, passing through the thermistor g(made of the same material). The change in the resistance of the two thermistors are noted which confirms the occurrence of the reaction.

In breast cancer, a surface protein HER2 is overexpressed. HERCEPTIN is a drug which is actually a monoclonal antibody which reacts with breast cells having overexpressed HER2. The reaction of the drug with the protein is an exothermic reaction and can be detected using a calorimetric transducer and hence cancer detection is possible.

Electrochemical Transducer

Electrochemical transducer reports changes in form of electrical signal which is directly proportional to the concentration of analyte.Binding of analyte results in ionic discharge, which can then be measured in the form of current or voltage, using suitable transducers.

Principle

Electrochemical reactions take place at electrode-electrolyte interfaces and provide a switch for electricity to flow between two phases of different conductivity, i.e. the electrode (electrons or holes are the charge carriers) and solid or liquid electrolyte (ions are the main charge carriers).

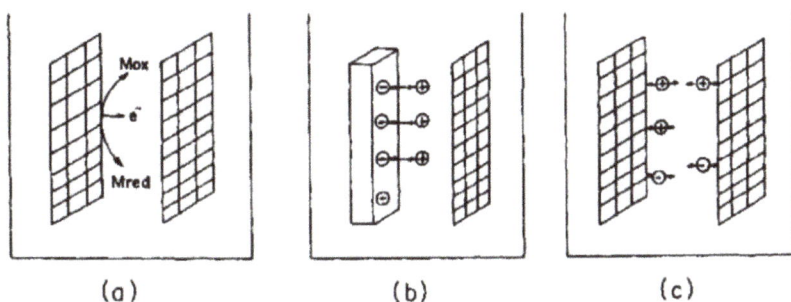

Figure 2. Principle of (a) amperometric, (b) potentiometric and (c) conductimetric biosensors; M_{red}, reduced mediator; M_{ox}, oxidised mediator; $\ominus$, anion; $\oplus$, cation.

Applications

Glucose Oxidase

$$\text{Glucose} + O_2 \rightarrow \text{Gluconic Acid} + H_2O_2$$

The product, H_2O_2,is oxidized at +650mV vsaAg/AgCl reference electrode.Thus, a potential of + 650mV is applied and the oxidation of H_2O_2 measured. This current is directly proportional to the concentration of glucose.

An electrochemical method for monitoring biotin–streptavidin interaction has been developed. This is based on the use of colloidal gold as an electrochemical label. Biotinylated albumin is adsorbed on the pretreated surface of a carbon paste electrode. This modified electrode is immersed in the colloidal gold–streptavidin labeled solution. Adsorptive voltametry is used to monitor colloidal gold bound to streptavidin. The analytical signal is highly reproducible sequential competitive assay.

Disposable immunochromatographic sensor for on-line quantitative determination of human serum albumin (HSA) has been developed. The sensor used conductimetric detection and 20 nm gold colloid particles modified with polyaniline (a conducting polymer) for signal generation. The reaction between the conjugate and analyte took place immediately and this complex was carried up into the next membrane that had the immobilized antibody. The second antigen–antibody reaction formed a sandwich-type immune complex at the electrode and polyaniline-bound colloidal gold generated a conductimetric signal.

A novel array-based electrical detection of DNA with nanoparticle probes has been developed that could be used to detect target DNA at concentrations as low as 500 fM with a point mutation selectivity factor of ~ 100,000:1.

The modification of electrochemical transducers with carbon nanotubes (CNTs) has recently attracted considerable attention in the field of DNA sensing technology.

Magnetic Transducers

Magnetic transducer makes use of the phenomenon of emf induction by change in the magnetic flux. The electrical signal hence produced provides the detection of the targeted element.

Magnetic transducers have their application in nanobiotechnology in detecting a specific cell. Nanoparticles containing magnetic material are engineered biotechnically and aptamers or antibodies which bind to the target cells are attached to the surface. The nanoparticles are introduced inside the system where they bind to the specific cells. Then, they are magnetized by passing an electric field. The magnetization leads to the flux change through the transducer which identifies the targeted cells through an electric signal. After that even the separation or the removal of such cells is possible through the application of a strong magnetic field.

GMR- giant magnetoresistance is a phenomenon seen in structures in alternating ferromagnetic and non magnetic material. It leads to a change in resistance if the magnetization in the alternating ferromagnetic films is parallel in one and anti parallel in the other. It is a highly sensitive method of detecting magnetic changes and is also used in the fabrication of transducers.

Telomeres are unique DNA–protein structures that contain noncoding long, repetitive sequences of TTAGGG and telomere-associated proteins. A novel nanosensor based on magnetic nanoparticles has been developed for rapid screens of telomerase activity in biological samples. The technique utilizes nanoparticles, which upon annealing with telomerase-associated TTAGGG repeats, switch their magnetic state, a phenomenon readily detectable by magnetic readers. High- throughput adaptation of the technique by magnetic resonance imaging allowed processing of hundreds of samples within tens of minutes at ultrahigh sensitivities. Together, these studies establish and validate a novel and powerful tool for rapidly sensing telomerase activity and provide the rationale for developing analogous magnetic nanoparticles for in vivo sensing. Since elevated telomerase levels are found in many malignancies, this technique offers provides access to an attractive target for therapeutic intervention and for diagnostic or prognostic purposes.

Piezoelectric Transducer

Piezoelectric transducer is the one that converts change in pressure or mass to an electrical field, thereby behaving like a sensor. It is made up of a piezoelectric material. Example of a piezoelectric material is quartz. Quartz is a sheet structure of silica with the molecular formula SiO_2. It is polar in nature. Any mechanical stress leads to a change in structure and hence the separation between positive and negative centres changes, causing a net electric field. If such a material is connected to a circuit, the mechanical stress will lead to an electrical signal, which in turn can be detected.

Piezoelectric transducers can be used as biosensors in a way that a molecular recognition element, say, an antibody can be made attached to a piezoelectric electrode. If the analyte binds to the antibody or any bulky molecule has an affinity towards the analyte which is already bonded to the antibody, the change in mass causes the electrode to experience sufficient pressure so as to produce an electrical field which in turn changes the resonating frequency of the crystal which can be noted and hence works as a beacon to show the binding of the analyte with the provided antibody.

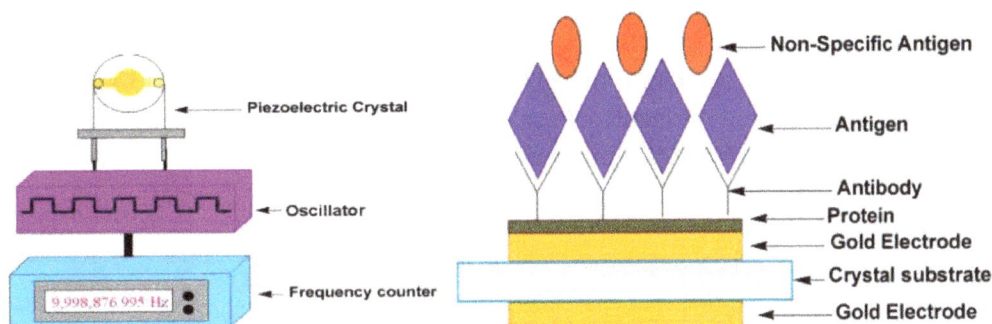

Optical Transducers

Optical transduction utilizes changes in optical properties such as phase, amplitude, and frequency, manifested because of the selective binding of an analyte with the bio-recognition element i.e. it converts chemical to light energy.

Principle

- Colorimetric for color: Measure change in light adsorbed as reactants are converted to products.

- Photometric for light intensity: Photon output for a luminescent or fluorescent process can be detected with photomultiplier tubes or photodiode systems

The governing thing is that enzymatic reactions alter the optical properties of some substances which allow them to emit light upon illumination in the form of some fluorescence, phosphorescence or chemiluminescence.

The SPR(Surface Plasmon Resonance) is an optical phenomenon due to a charge density oscillation at the interface of a metal and a dielectric, which has dielectric constants of opposite signs. For example, a thin layer of gold on a high refractive index glass surface can absorb laser light, producing electron waves (surface plasmons) on the gold surface. This occurs only at a specific angle and wavelength of incident light and is highly dependent on the surface of the gold, such that binding of a target analyte to a receptor changes the resonant frequency which produces a measurable signal.

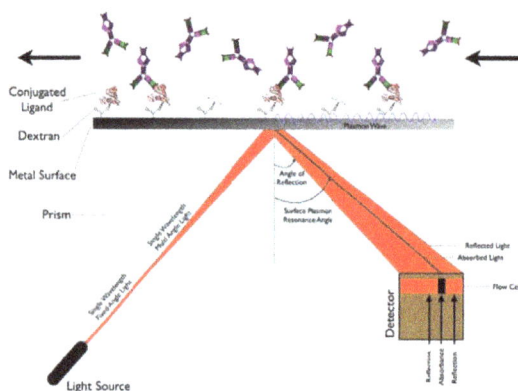

Fluorescence is a molecular absorption of light at one wavelength and its instantaneous emission of at longer wavelengths. Some molecules fluoresce naturally and others such as DNA can be modified for fluorescence detection by attachment of special fluorescent dyes.

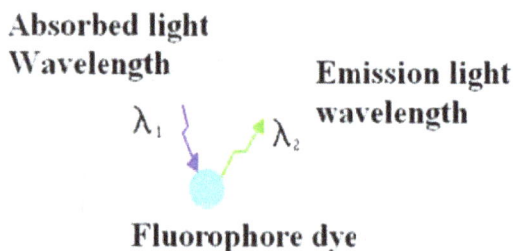

Applications

DNA Sensors: Genetic monitoring, disease, Immunosensors: HIV, Hepatitis, other viral disease, drug testing, environmental monitoring, Cell-based Sensors: functional sensors, drug testing, Point-of-care sensors: blood, urine, electrolytes, gases, steroids,drugs, hormones, and proteins, Bacteria Sensors: (E-coli, streptococcus, other): food industry, medicine, environmental. Enzyme sensors: diabetics, drug testing.

Gold nanoparticles have been used as a new class of universal fluorescence quenchers to develop an optical biosensor for recognizing and detecting specific DNA sequences.

Quantam dots are used as inorganic fluorophores. QDs have fairly broad excitation spectra–from ultraviolet to red–that can be tuned depending on their size and composition. They are being used in virus tagging and cancer cells imaging for diagnostics purposes.

A most promising transducer involving luminescence uses firefly luciferase (Photinus-luciferin 4-monooxygenase (ATP-hydrolysing), EC 1.13.12.7) to detect the presence of bacteria in food or clinical samples. Bacteria are specifically lysed and the ATP released (roughly proportional to the number of bacteria present) reacted with D-luciferin and oxygen in a reaction which produces yellow light in high quantum yield.

$$\overset{\text{luciferase}}{\text{ATP+D-luciferin}+O_2 \rightarrow \text{oxyluciferin+AMP+pyrophosphate+}CO_2+\text{light(562nm)}}$$

Mechanical Transducer

Mechanical Transducer is one which converts mechanical energy into electrical energy.

Mechanical Transducers are usually cantilever shaped. Smaller size of mechanical transducers produces outstanding mass resolution in a single atom which is an advantage from quartz crystal.

In this case, the bio molecules form a monolayer on the upper side of the micro cantilever that generates surface stress producing static mechanical bending with curvature which causes cantilever to oscillate with respect to the equilibrium position.

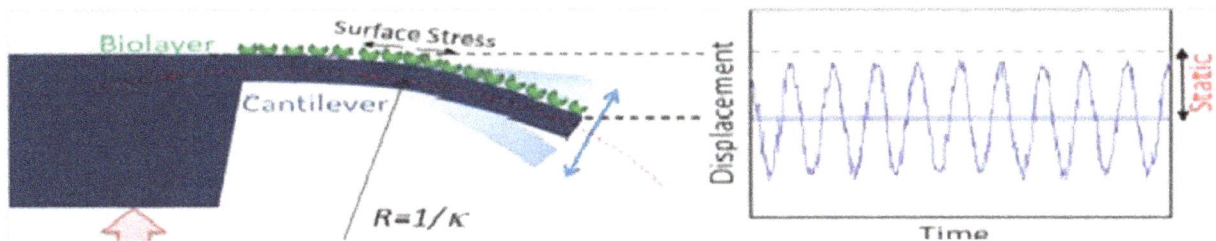

Applications

Biosensors based on mechanical transducers can specifically detect single-base mismatches in oligonucleotide hybridization. In these assays nucleic acids are immobilized on a side of a cantilever (active side). Exposure of the cantilever to a sample containing complementary nucleic acid gives rise to a cantilever bending (deflection) of a few nanometers. Deflection can be measured by an optical system in which a laser beam reflects on cantilever.

Various types of transducers have been discussed after thorough study is made. Their application in bio-nanotechnology has also been described. It is concluded that nano transducers provide much better results over conventional sensors. Further scope is their controlled production and better detecting techniques which can enhance the use these in medical purposes. There seems to be bright future for detection once the appropriate technology is developed.

Biosensor

A biosensor is an analytical device, used for the detection of an analyte, that combines a biological component with a physicochemical detector. The *sensitive biological element* (e.g. tissue, microorganisms, organelles, cell receptors, enzymes, antibodies, nucleic acids, etc.) is a biologically derived material or biomimetic component that interacts (binds or recognizes) with the analyte under study. The biologically sensitive elements can also be created by biological engineering. The *transducer* or the *detector element* (works in a physicochemical way; optical, piezoelectric, electrochemical, etc.) transforms the signal resulting from the interaction of the analyte with the biological element into another signal (i.e., transduces) that can be more easily measured and quantified. The biosensor reader device with the associated electronics or signal processors that are primarily responsible for the display of the results in a user-friendly way. This sometimes accounts for the most expensive part of the sensor device, however it is possible to generate a user friendly display that includes transducer and sensitive element (holographic sensor). The readers are usually custom-designed and manufactured to suit the different working principles of biosensors.

Biosensor System

A biosensor typically consists of a bio-recognition site, biotransducer component, and electronic system which include a signal amplifier, processor, and display. Transducers and electronics can be combined, e.g., in CMOS-based microsensor systems. The recognition component, often called a bioreceptor, uses biomolecules from organisms or receptors modeled after biological systems to interact with the analyte of interest. This interaction is measured by the biotransducer which outputs a measurable signal proportional to the presence of the target analyte in the sample. The general aim of the design of a biosensor is to enable quick, convenient testing at the point of concern or care where the sample was procured.

Bioreceptors

In a biosensor, the bioreceptor is designed to interact with the specific analyte of interest to produce an effect measurable by the transducer. High selectivity for the analyte among a matrix of other chemical or biological components is a key requirement of the bioreceptor. While the type of biomolecule used can vary widely, biosensors can be classified according to common types bioreceptor interactions involving: anitbody/antigen, enzymes/ligands, nucleic acids/DNA, cellular structures/cells, or biomimetic materials.

Antibody/Antigen Interactions

An immunosensor utilizes the very specific binding affinity of antibodies for a specific compound or antigen. The specific nature of the antibody-antigen interaction is analogous to a lock and key fit in that the antigen will only bind to the antibody if it has the correct conformation. Binding events result in a physicochemical change that in combination with a tracer, such as a fluorescent molecules, enzymes, or radioisotopes, can generate a signal. There are limitations with using antibodies in sensors: 1.The antibody binding capacity is strongly dependent on assay conditions (e.g. pH and temperature) and 2. The antibody-antigen interaction is generally irreversible. However, it has been shown that binding can be disrupted by chaotropic reagents, organic solvents, or even ultrasonic radiation.

Artificial Binding Proteins

The use of antibodies as the bio-recognition component of biosensors has several drawbacks. They have high molecular weights and limited stability, contain essential disulfide bonds and are expensive to produce. In one approach to overcome these limitations, recombinant binding fragments (Fab, Fv or scFv) or domains (VH, VHH) of antibodies have been engineered. In another approach, small protein scaffolds with favorable biophysical properties have been engineered to generate artificial families of Antigen Binding Proteins (AgBP), capable of specific binding to different target proteins while retaining the favorable properties of the parent molecule. The elements of the family that specifically bind to a given target antigen, are often selected in vitro by display techniques: phage display, ribosome display, yeast display or mRNA display. The artificial binding proteins are much smaller than antibodies (usually less than 100 amino-acid residues), have a strong stability, lack disulfide bonds and can be expressed in high yield in reducing cellular environments like the bacterial cytoplasm, contrary to antibodies and their derivatives. They are thus especially suitable to create biosensors.

Enzymatic Interactions

The specific binding capabilities and catalytic activity of enzymes make them popular bioreceptors. Analyte recognition is enabled through several possible mechanisms: 1) the enzyme converting the analyte into a product that is sensor-detectable, 2) detecting enzyme inhibition or activation by the analyte, or 3) monitoring modification of enzyme properties resulting from interaction with the analyte. The main reasons for the common use of enzymes in biosensors are: 1) ability to catalyze a large number of reactions; 2) potential to detect a group of analytes (substrates, products, inhibitors, and modulators of the catalytic activity); and 3) suitability with several different transduction methods for detecting the analyte. Notably, since enzymes are not consumed in reactions, the biosensor can easily be used continuously. The catalytic activity of enzymes also allows lower limits of detection compared to common binding techniques. However, the sensor's lifetime is limited by the stability of the enzyme.

Affinity Binding Receptors

Antibodies have a high binding constant in excess of 10^8 L/mol, which stands for a nearly irreversible association once the antigen-antibody couple has formed. For certain analyte molecules like glucose affinity binding proteins exist that bind their ligand with a high specificity like an antibody, but with a much smaller binding constant on the order of 10^2 to 10^4 L/mol. The association between analyte and receptor then is of reversible nature and next to the couple between both also their free molecules occur in a measurable concentration. In case of glucose, for instance, concanavalin A may function as affinity receptor exhibiting a binding constant of 4×10^2 L/mol. The use of affinity binding receptors for purposes of biosensing has been proposed by Schultz and Sims in 1979 and was subsequently configured into a fluorescent assay for measuring glucose in the relevant physiological range between 4.4 and 6.1 mmol/L. The sensor principle has the advantage that it does not consume the analyte in a chemical reaction as is occurs in enzymatic assays.

Nucleic Acid Interactions

Biosensors that employ nucleic acid interactions can be referred to as genosensors. The recognition process is based on the principle of complementary base pairing, adenine:thymine and cytosine:guanine in DNA. If the target nucleic acid sequence is known,complementary sequences can be synthesized, labeled, and then immobilized on the sensor. The hybridization probes can then base pair with the target sequences, generating an optical signal. The favored transduction principle employed in this type of sensor has been optical detection.

Epigenetics

It has been proposed that properly optimized integrated optical resonators can be exploited for detecting epigenetic modifications (e.g. DNA methylation, histone post-translational modifications) in body fluids from patients affected by cancer or other diseases. Photonic biosensors with ultra-sensitivity are nowadays being developed at a research level to easily detect cancerous cells within the patient's urine. Different research projects aim to develop new portable devices that uses cheap, environmentally friendly, disposable cartridges that require only simple handling with no need of further processing, washing, or manipulation by expert technicians.

Organelles

Organelles form separate compartments inside cells and usually perform function independently. Different kinds of organelles have various metabolic pathways and contain enzymes to fulfill its function. Commonly used organelles include lysosome, chloroplast and mitochondria. The spatial-temporal distribution pattern of calcium is closed related to ubiquitous signaling pathway. Mitochondria actively participate in the metabolism of calcium ions to control the function and also modulate the calcium related signaling pathways. Experiments have proved that mitochondria have the ability to respond to high calcium concentration generated in the proximity by opening the calcium channel. In this way, mitochondria can be used to detect the calcium concentration in medium and the detection is very sensitive due to high spatial resolution. Another application of mitochondria is used for detection of water pollution. Detergent compounds' toxicity will damage the cell and subcellular structure including mitochondria. The detergents will cause a swelling effect which could be measured by an absorbance change. Experiment data shows the change rate is proportional to the detergent concentration, providing a high standard for detection accuracy.

Cells

Cells are often used in bioreceptors because they are sensitive to surrounding environment and they can respond to all kinds of stimulants. Cells tend to attach to the surface so they can be easily immobilized. Compared to organelles they remain active for longer period and the reproducibility makes them reusable. They are commonly used to detect global parameter like stress condition, toxicity and organic derivatives. They can also be used to monitor the treatment effect of drugs. One application is to use cells to determine herbicides which are main aquatic contaminant. Microalgae are entrapped on a quartz microfiber and the chlorophyll fluorescence modified by herbicides is collected at the tip of an optical fiber bundle and transmitted to a fluorimeter. The algae are continuously cultured to get optimized measurement. Results show that detection limit of certain herbicide can reach sub-ppb concentration level. Some cells can also be used to monitor the microbial corrosion. Pseudomonas sp. is isolated form corroded material surface and immobilized on acetylcellulose membrane. The respiration activity is determined by measuring oxygen consumption. There is linear relationship between the current generated and the concentration of sulfuric acid. The response time is related to the loading of cells and surrounding environments and can be controlled to no more than 5min.

Tissue

Tissues are used for biosensor for the abundance of enzymes existed. Advantages of tissues as biosensors include the following: 1)easier to immobilize compared to cells and organelles 2)the higher activity and stability from maintain enzymes in natural environment 3)the availability and low price 4)the avoidance of tedious work of extraction, centrifuge and purification of enzymes 5)necessary cofactors for enzyme to function exists 6)the diversity providing a wide range of choice concerning different objectives. There also exists some disadvantages of tissues like the lack of specificity due to the interference of other enzymes and longer response time due to transport barrier.

Surface Attachment of the Biological Elements

An important part in a biosensor is to attach the biological elements (small molecules/protein/cells) to the surface of the sensor (be it metal, polymer or glass). The simplest way is to functionalize the surface in order to coat it with the biological elements. This can be done by polylysine, aminosilane, epoxysilane or nitrocellulose in the case of silicon chips/silica glass. Subsequently, the bound biological agent may be for example fixed by Layer by layer depositation of alternatively charged polymer coatings. Alternatively three-dimensional lattices (hydrogel/xerogel) can be used to chemically or physically entrap these (where by chemically entraped it is meant that the biological element is kept in place by a strong bond, while physically they are kept in place being unable to pass through the pores of the gel matrix). The most commonly used hydrogel is sol-gel, a glassy silica generated by polymerization of silicate monomers (added as tetra alkyl orthosilicates, such as TMOS or TEOS) in the presence of the biological elements (along with other stabilizing polymers, such as PEG) in the case of physical entrapment. Another group of hydrogels, which set under conditions suitable for cells or protein, are acrylate hydrogel, which polymerize upon radical initiation. One type of radical initiator is a peroxide radical, typically generated by combining a persulfate with TEMED (Polyacrylamide gel are also commonly used for protein electrophoresis), alternatively light can be used in combination with a photoinitiator, such as DMPA (2,2-dimethoxy-2-phenylacetophenone). Smart materials that mimic the biological components of a sensor can also be classified as biosensors using only the active or catalytic site or analogous configurations of a biomolecule.

Biotransducer

Biosensors can be classified by their biotransducer type. The most common types of biotransducers used in biosensors are 1) electrochemical biosensors, 2) optical biosensors, 3)electronic biosensors, 4)piezoelectric biosensors, 5) gravimetric biosensors, 6) pyroelectric biosensors.

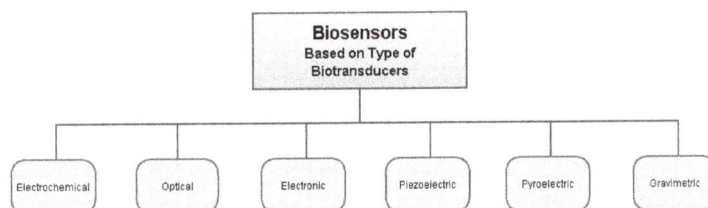

Classification of Biosensors based on type of biotransducer

Electrochemical

Electrochemical biosensors are normally based on enzymatic catalysis of a reaction that produces or consumes electrons (such enzymes are rightly called redox enzymes). The sensor substrate usually contains three electrodes; a reference electrode, a working electrode and a counter electrode. The target analyte is involved in the reaction that takes place on the active electrode surface, and the reaction may cause either electron transfer across the double layer (producing a current) or can contribute to the double layer potential (producing a voltage). We can either measure the current (rate of flow of electrons is now proportional to the analyte concentration) at a fixed potential or the potential can be measured at zero current (this gives a logarithmic response). Note that potential of the working or active electrode is space charge sensitive and this is often used. Further, the label-free and direct electrical detection of small peptides and proteins is possible by their intrinsic charges using biofunctionalized ion-sensitive field-effect transistors.

Another example, the potentiometric biosensor, (potential produced at zero current) gives a logarithmic response with a high dynamic range. Such biosensors are often made by screen printing the electrode patterns on a plastic substrate, coated with a conducting polymer and then some protein (enzyme or antibody) is attached. They have only two electrodes and are extremely sensitive and robust. They enable the detection of analytes at levels previously only achievable by HPLC and LC/MS and without rigorous sample preparation. All biosensors usually involve minimal sample preparation as the biological sensing component is highly selective for the analyte concerned. The signal is produced by electrochemical and physical changes in the conducting polymer layer due to changes occurring at the surface of the sensor. Such changes can be attributed to ionic strength, pH, hydration and redox reactions, the latter due to the enzyme label turning over a substrate. Field effect transistors, in which the gate region has been modified with an enzyme or antibody, can also detect very low concentrations of various analytes as the binding of the analyte to the gate region of the FET cause a change in the drain-source current.

Ion Channel Switch

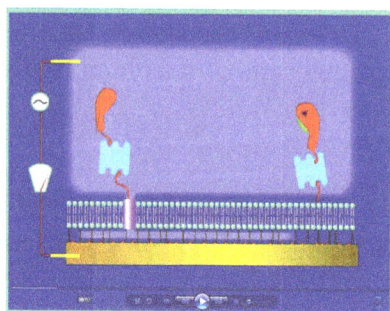

ICS – channel open   ICS – channel closed

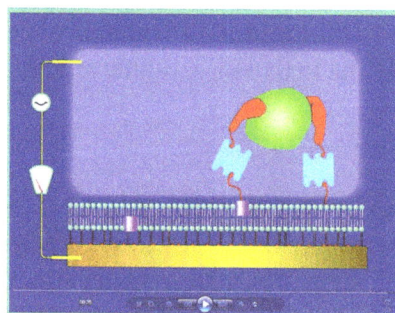

The use of ion channels has been shown to offer highly sensitive detection of target biological molecules. By embedding the ion channels in supported or tethered bilayer membranes (t-BLM) attached to a gold electrode, an electrical circuit is created. Capture molecules such as antibodies can be bound to the ion channel so that the binding of the target molecule controls the ion flow through the channel. This results in a measurable change in the electrical conduction which is proportional to the concentration of the target.

An ion channel switch (ICS) biosensor can be created using gramicidin, a dimeric peptide channel, in a tethered bilayer membrane. One peptide of gramicidin, with attached antibody, is mobile and one is fixed. Breaking the dimer stops the ionic current through the membrane. The magnitude of the change in electrical signal is greatly increased by separating the membrane from the metal surface using a hydrophilic spacer.

Quantitative detection of an extensive class of target species, including proteins, bacteria, drug and toxins has been demonstrated using different membrane and capture configurations.

Reagentless Fluorescent Biosensor

A reagentless biosensor can monitor a target analyte in a complex biological mixture without additional reagent. Therefore, it can function continuously if immobilized on a solid support. A fluorescent biosensor reacts to the interaction with its target analyte by a change of its fluorescence properties. A Reagentless Fluorescent biosensor (RF biosensor) can be obtained by integrating a biological receptor, which is directed against the target analyte, and a solvatochromic fluorophore, whose emission properties are sensitive to the nature of its local environment, in a single macromolecule. The fluorophore transduces the recognition event into a measurable optical signal. The use of extrinsic fluorophores, whose emission properties differ widely from those of the intrinsic fluorophores of proteins, tryptophan and tyrosine, enables one to immediately detect and quantify the analyte in complex biological mixtures. The integration of the fluorophore must be done in a site where it is sensitive to the binding of the analyte without perturbing the affinity of the receptor.

Antibodies and artificial families of Antigen Binding Proteins (AgBP) are well suited to provide the recognition module of RF biosensors since they can be directed against any antigen. A general approach to integrate a solvatochromic fluorophore in an AgBP when the atomic structure of the complex with its antigen is known, and thus transform it into a RF biosensor, has been described. A residue of the AgBP is identified in the neighborhood of the antigen in their complex. This residue is changed into a cysteine by site-directed mutagenesis. The fluorophore is chemically coupled to the mutant cysteine. When the design is successful, the coupled fluorophore does not prevent the binding of the antigen, this binding shields the fluorophore from the solvent, and it can be detected by a change of fluorescence. This strategy is also valid for antibody fragments.

However, in the absence of specific structural data, other strategies must be applied. Antibodies and artificial families of AgBPs are constituted by a set of hypervariable (or randomized) residue positions, located in a unique sub-region of the protein, and supported by a constant polypeptide scaffold. The residues that form the binding site for a given antigen, are selected among the hypervariable residues. It is possible to transform any AgBP of these families into a RF biosensor, specific of the target antigen, simply by coupling a solvatochromic fluorophore to one of the hypervariable residues that have little or no importance for the interaction with the antigen, after changing this residue into cysteine by mutagenesis. More specifically, the strategy consists in individually changing the residues of the hypervariable positions into cysteine at the genetic level, in chemically coupling a solvatochromic fluorophore with the mutant cysteine, and then in keeping the resulting conjugates that have the highest sensitivity (a parameter that involves both affinity and variation of fluorescence signal). This approach is also valid for families of antibody fragments.

A posteriori studies have shown that the best reagentless fluorescent biosensors are obtained when the fluorophore does not make non-covalent interactions with the surface of the bioreceptor, which would increase the background signal, and when it interacts with a binding pocket at the surface of the target antigen. The RF biosensors that are obtained by the above methods, can function and detect target analytes inside living cells.

Others

Piezoelectric sensors utilise crystals which undergo an elastic deformation when an electrical potential is applied to them. An alternating potential (A.C.) produces a standing wave in the crystal at a characteristic frequency. This frequency is highly dependent on the elastic properties of the crystal, such that if a crystal is coated with a biological recognition element the binding of a (large) target analyte to a receptor will produce a change in the resonance frequency, which gives a binding signal. In a mode that uses surface acoustic waves (SAW), the sensitivity is greatly increased. This is a specialised application of the Quartz crystal microbalance as a biosensor

Thermometric and magnetic based biosensors are rare.

Placement of Biosensors

The appropriate placement of biosensors depends on their field of application, which may roughly be divided into biotechnology, agriculture, food technology and biomedicine.

In biotechnology, analysis of the chemical composition of cultivation broth can be conducted in-line, on-line, at-line and off-line. As outlined by the US Food and Drug Administration (FDA) the sample is not removed from the process stream for in-line sensors, while it is diverted from the manufacturing process for on-line measurements. For at-line sensors the sample may be removed and analyzed in close proximity to the process stream. An example of the latter is the monitoring of lactose in a dairy processing plant. Off-line biosensors compare to bioanalytical techniques that are not operating in the field, but in the laboratory. These techniques are mainly used in agriculture, food technology and biomedicine.

Medical biosensor implant for glucose monitoring in subcutaneous tissue (59x45x8 mm). Electronic components like microcontroller, radio chip etc. are hermetically enclosed in a Ti casing, while ring-shaped antenna and top-most sensor probe are moulded into the transparent epoxy header.

In medical applications biosensors are generally categorized as in vitro and in vivo systems. An in vitro biosensor measurement takes place in a test tube, a culture dish, a microtiter plate or elsewhere outside a living organism. The sensor uses a bioreceptor and transducer as outlined above. An example of an in vitro biosensor is an enzyme-conductimetric biosensor for blood glucose monitoring. There is a challenge to create a biosensor that operates by the principle of Point-of-care testing, i.e. at the location where the test is needed. The elimination of lab testing can save time and money. An application of a POCT biosensor can be for the testing of HI virus virus in areas, where it is difficult for patients to be tested. A biosensor can be sent directly to the location and a quick and easy test can be used.

An in vivo biosensor is an implantable device that operates inside the body. Of course, biosensor implants have to fulfill the strict regulations on sterilization in order to avoid an initial inflammatory response after implantation. The second concern relates to the long-term biocompatibility, i.e. the unharmful interaction with the body environment during the intended period of use. Another issue that arises is failure. If there is failure, the device must be removed and replaced, causing additional surgery. An example for application of an in vivo biosensor would be the insulin monitoring within the body, which is not available yet.

Most advanced biosensor implants have been developed for the continuous monitoring of glucose. The figure displays a device, for which a Ti casing and a battery as established for cardiovascular implants like pacemakers and defibrillators is used. Its size is determined by the battery as required for a lifetime of one year. Measured glucose data will be transmitted wirelessly out of the body within the MICS 402-405 MHz band as approved for medical implants.

Nowadays biosensors are being integrated into mobile phone systems, making them user-friendly and accessible to a large amount of users.

Applications

There are many potential applications of biosensors of various types. The main requirements for a biosensor approach to be valuable in terms of research and commercial applications are the identification of a target molecule, availability of a suitable biological recognition element, and the potential for disposable portable detection systems to be preferred to sensitive laboratory-based techniques in some situations. Some examples are glucose monitoring in diabetes patients, other medical health related targets, environmental applications e.g. the detection of pesticides and river water contaminants such as heavy metal ions, remote sensing of airborne bacteria e.g. in counter-bioterrorist activities, remote sensing of water quality in coastal waters by describing online different aspects of clam ethology (biological rhythms, growth rates, spawning or death records) in groups of abandoned bivalves around the world, detection of pathogens, determining levels of toxic substances before and after bioremediation, detection and determining of organophosphate, routine analytical measurement of folic acid, biotin, vitamin B12 and pantothenic acid as an alternative to microbiological assay, determination of drug residues in food, such as antibiotics and growth promoters, particularly meat and honey, drug discovery and evaluation of biological activity of new compounds, protein engineering in biosensors, and detection of toxic metabolites such as mycotoxins.

A common example of a commercial biosensor is the blood glucose biosensor, which uses the enzyme glucose oxidase to break blood glucose down. In doing so it first oxidizes glucose and uses two electrons to reduce the FAD (a component of the enzyme) to FADH2. This in turn is oxidized by the electrode in a number of steps. The resulting current is a measure of the concentration of glucose. In this case, the electrode is the transducer and the enzyme is the biologically active component.

A canary in a cage, as used by miners to warn of gas, could be considered a biosensor. Many of today's biosensor applications are similar, in that they use organisms which respond to toxic substances at a much lower concentrations than humans can detect to warn of their presence. Such devices can be used in environmental monitoring, trace gas detection and in water treatment facilities.

Many optical biosensors are based on the phenomenon of surface plasmon resonance (SPR) techniques. This utilises a property of and other materials; specifically that a thin layer of gold on a high refractive index glass surface can absorb laser light, producing electron waves (surface plasmons) on the gold surface. This occurs only at a specific angle and wavelength of incident light and is highly dependent on the surface of the gold, such that binding of a target analyte to a receptor on the gold surface produces a measurable signal.

Surface plasmon resonance sensors operate using a sensor chip consisting of a plastic cassette supporting a glass plate, one side of which is coated with a microscopic layer of gold. This side contacts the optical detection apparatus of the instrument. The opposite side is then contacted with a microfluidic flow system. The contact with the flow system creates channels across which reagents can be passed in solution. This side of the glass sensor chip can be modified in a number of ways, to allow easy attachment of molecules of interest. Normally it is coated in carboxymethyl dextran or similar compound.

The refractive index at the flow side of the chip surface has a direct influence on the behavior of the light reflected off the gold side. Binding to the flow side of the chip has an effect on the refractive index and in this way biological interactions can be measured to a high degree of sensitivity with some sort of energy. The refractive index of the medium near the surface changes when biomolecules attach to the surface, and the SPR angle varies as a function of this change.

Light of a fixed wavelength is reflected off the gold side of the chip at the angle of total internal reflection, and detected inside the instrument. The angle of incident light is varied in order to match the evanescent wave propagation rate with the propagation rate of the surface plasmon plaritons. This induces the evanescent wave to penetrate through the glass plate and some distance into the liquid flowing over the surface.

Other optical biosensors are mainly based on changes in absorbance or fluorescence of an appropriate indicator compound and do not need a total internal reflection geometry. For example, a fully operational prototype device detecting casein in milk has been fabricated. The device is based on detecting changes in absorption of a gold layer. A widely used research tool, the micro-array, can also be considered a biosensor.

Biological biosensors often incorporate a genetically modified form of a native protein or enzyme. The protein is configured to detect a specific analyte and the ensuing signal is read by a detection instrument such as a fluorometer or luminometer. An example of a recently developed biosensor is one for detecting cytosolic concentration of the analyte cAMP (cyclic adenosine monophosphate), a second messenger involved in cellular signaling triggered by ligands interacting with receptors on the cell membrane. Similar systems have been created to study cellular responses to native ligands or xenobiotics (toxins or small molecule inhibitors). Such "assays" are commonly used in drug discovery development by pharmaceutical and biotechnology companies. Most cAMP assays in current use require lysis of the cells prior to measurement of cAMP. A live-cell biosensor for cAMP can be used in non-lysed cells with the additional advantage of multiple reads to study the kinetics of receptor response.

Nanobiosensors use an immobilized bioreceptor probe that is selective for target analyte molecules. Nanomaterials are exquisitely sensitive chemical and biological sensors. Nanoscale materials demonstrate unique properties. Their large surface area to volume ratio can achieve rapid and low cost reactions, using a variety of designs.

Other evanescent wave biosensors have been commercialised using waveguides where the propagation constant through the waveguide is changed by the absorption of molecules to the waveguide surface. One such example, dual polarisation interferometry uses a buried waveguide as a reference against which the change in propagation constant is measured. Other configurations such as the Mach–Zehnder have reference arms lithographically defined on a substrate. Higher levels of integration can be achieved using resonator geometries where the resonant frequency of a ring resonator changes when molecules are absorbed.

Recently, arrays of many different detector molecules have been applied in so called electronic nose devices, where the pattern of response from the detectors is used to fingerprint a substance. In the Wasp Hound odor-detector, the mechanical element is a video camera and the biological element is five parasitic wasps who have been conditioned to swarm in response to the presence of a specific chemical. Current commercial electronic noses, however, do not use biological elements.

Glucose Monitoring

Commercially available gluocose monitors rely on amperometric sensing of glucose by means of glucose oxidase, which oxidises glucose producing hydrogen peroxide which is detected by the electrode. To overcome the limitation of amperometric sensors, a flurry of research is present into novel sensing methods, such as fluorescent glucose biosensors.

Interferometric Reflectance Imaging Sensor

The interferometric reflectance imaging sensor (IRIS) is based on the principles of optical interference and consists of a silicon-silicon oxide substrate, standard optics, and low-powered coherent LEDs. When light is illuminated through a low magnification objective onto the layered silicon-silicon oxide substrate, an interferometric signature is produced. As biomass, which has a similar index of refraction as silicon oxide, accumulates on the substrate surface, a change in the interferometric signature occurs and the change can be correlated to a quantifiable mass. *Daaboul et al.* used IRIS to yield a label-free sensitivity of approximately 19 ng/mL. *Ahn et al.* improved the sensitivity of IRIS through a mass tagging technique.

Since initial publication, IRIS has been adapted to perform various functions. First, IRIS integrated a fluorescence imaging capability into the interferometric imaging instrument as a potential way to address fluorescence protein microarray variability. Briefly, the variation in fluorescence microarrays mainly derives from inconsistent protein immobilization on surfaces and may cause misdiagnoses in allergy microarrays. To correct from any variation in protein immobilization, data acquired in the fluorescence modality is then normalized by the data acquired in the label-free modality. IRIS has also been adapted to perform single nanoparticle counting by simply switching the low magnification objective used for label-free biomass quantification to a higher objective magnification. This modality enables size discrimination in complex human biological samples. *Monroe et al..* used IRIS to quantify protein levels spiked into human whole blood and serum and determined allergen sensitization in characterized human blood samples using zero sample processing. Other practical uses of this device include virus and pathogen detection.

Food Analysis

There are several applications of biosensors in food analysis. In the food industry, optics coated with antibodies are commonly used to detect pathogens and food toxins. Commonly, the light system in these biosensors is fluorescence, since this type of optical measurement can greatly amplify the signal.

A range of immuno- and ligand-binding assays for the detection and measurement of small molecules such as water-soluble vitamins and chemical contaminants (drug residues) such as sulfonamides and Beta-agonists have been developed for use on SPR based sensor systems, often adapted from existing ELISA or other immunological assay. These are in widespread use across the food industry.

DNA Biosensors

In the future, DNA will find use as a versatile material from which scientists can craft biosensors. DNA biosensors can theoretically be used for medical diagnostics, forensic science, agriculture, or even environmental clean-up efforts. No external monitoring is needed for DNA-based sensing devises. This is a significant advantage. DNA biosensors are complicated mini-machines—consisting of sensing elements, micro lasers, and a signal generator. At the heart of DNA biosensor function is the fact that two strands of DNA stick to each other by virtue of chemical attractive forces. On such a sensor, only an exact fit—that is, two strands that match up at every nucleotide position—gives rise to a fluorescent signal (a glow) that is then transmitted to a signal generator.

Microbial Biosensors

Using biological engineering researchers have created many microbial biosensors. An example is the arsenic biosensor. To detect arsenic they use the Ars operon. Using bacteria, researchers can detect pollutants in samples.

Ozone Biosensors

Because ozone filters out harmful ultraviolet radiation, the discovery of holes in the ozone layer of the earth's atmosphere has raised concern about how much ultraviolet light reaches the earth's surface. Of particular concern are the questions of how deeply into sea water ultraviolet radiation penetrates and how it affects marine organisms, especially plankton (floating microorganisms) and viruses that attack plankton. Plankton form the base of the marine food chains and are believed to affect our planet's temperature and weather by uptake of CO_2 for photosynthesis.

Deneb Karentz, a researcher at the Laboratory of Radio-biology and Environmental Health (University of California in San Francisco) has devised a simple method for measuring ultraviolet penetration and intensity. Working in the Antarctic Ocean, she submerfed to various depths thin plastic bags containing special strains of *E. coli* that are almost totally unable to repair ultraviolet radiation damage to their DNA. Bacterial death rates in these bags were compared with rates in unexposed control bags of the same organism. The bacterial "biosensors" revealed constant significant ultraviolet damage at depths of 10 m and frequently at 20 and 30 m. Karentz plans additional studies of how ultraviolet may affect seasonal plankton blooms (growth spurts) in the oceans.

Metastatic Cancer Cell Biosensors

Metastasis is the spread of cancer from one part of the body to another via either the circulatory system or lymphatic system. Unlike radiology imaging tests (mammograms), which send forms of energy (x-rays, magnetic fields, etc.) through the body to only take interior pictures, biosensors have the potential to directly test the malignant power of the tumor. The combination of a biological and detector element allows for a small sample requirement, a compact design, rapid signals, rapid detection, high selectivity and high sensitivity for the analyte being studied. Compared to the usual radiology imaging tests biosensors have the advantage of not only finding out how far the cancer has spread and checking if treatment is effective, but also are cheaper, more efficient (in time, cost and productivity) ways to assess metastaticity in early stages of cancer.

Biological engineering researchers have created oncological biosensors for breast cancer. Breast cancer is the leading common cancer among women worldwide. An example would be a transferrin- quartz crystal microbalance (QCM). As a biosensor, quartz crystal microbalances produce oscillations in the frequency of the crystal's standing wave from an alternating potential to detect nano-gram mass changes. These biosensors are specifically designed to interact and have high selectivity for receptors on cell (cancerous and normal) surfaces. Ideally this provides a quantitative detection of cells with this receptor per surface area instead of a qualitative picture detection given by mammograms.

Seda Atay, a biotechnology researcher at Hacettepe University, experimentally observed this specificity and selectivity between a QCM and MDA-MB 231 breast cells, MCF 7 cells, and starved MDA-MB 231 cells in vitro. With other researchers she devised a method of washing these different metastatic leveled cells over the sensors to measure mass shifts due to different quantities of transferrin receptors. Particularly, the metastatic power of breast cancer cells can be determined by Quartz crystal microbalances with nanoparticles and transferrin that would potentially attach to transferrin receptors on cancer cell surfaces. There is very high selectivity for transferrin receptors because they are over-expressed in cancer cells. If cells have high expression of transferrin receptors, which shows their high metastatic power, they have higher affinity and bind more to the QCM that measures the increase in mass. Depending on the magnitude of the nano-gram mass change, the metastatic power can be determined.

Purpose of Nanoparticles

When nanoparticles are attached to the QCM surface their simplicity, variability in shape, high surface area, physicochemical malleability, and optional attachment of metals enables for different properties, a change in responses, selectivities and specificities. This transducer's characteristics in combination with nanoparticles with large surface area to volume ratios makes it a perfect biosensor to particularly determine the metastatic power and malignancy of cancer cells. In Seda

Atay's study, determining the metastatic power of in vitro breast cancer exactly required 58 nm sized Poly(2-hydroxyethyl methacrylate) (PHEMA) nanoparticles with a surface area of 1899 m2g-1 to effectively adsorb the cells to the QCM surface.

Quantum Dot

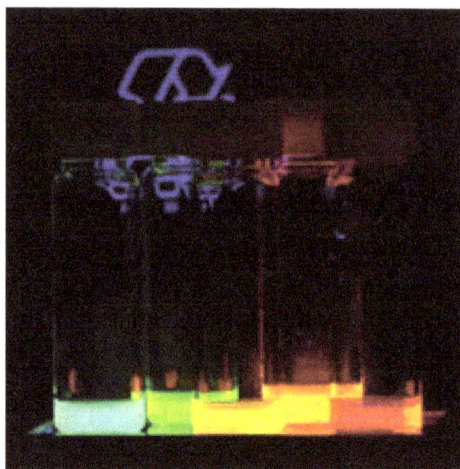

Colloidal quantum dots irradiated with a UV light. Different sized quantum dots emit different color light due to quantum confinement.

Quantum dots (QD) are very small semiconductor particles, only several nanometres in size, so small that their optical and electronic properties differ from those of larger particles. They are a central theme in nanotechnology. Many types of quantum dot will emit light of specific frequencies if electricity or light is applied to them, and these frequencies can be precisely tuned by changing the dots' size, shape and material, giving rise to many applications.

In the language of materials science, nanoscale semiconductor materials tightly confine either electrons or electron holes. Quantum dots are also sometimes referred to as artificial atoms, a term that emphasizes that a quantum dot is a single object with bound, discrete electronic states, as is the case with naturally occurring atoms or molecules.

Quantum dots exhibit properties that are intermediate between those of bulk semiconductors and those of discrete molecules. Their optoelectronic properties change as a function of both size and shape. Larger QDs (radius of 5–6 nm, for example) emit longer wavelengths resulting in emission colors such as orange or red. Smaller QDs (radius of 2–3 nm, for example) emit shorter wavelengths resulting in colors like blue and green, although the specific colors and sizes vary depending on the exact composition of the QD.

Because of their highly tunable properties, QDs are of wide interest. Potential applications include transistors, solar cells, LEDs, diode lasers and second-harmonic generation, quantum computing, and medical imaging. Additionally, their small size allows for QDs to be suspended in solution which leads to possible uses in inkjet printing and spin-coating. These processing techniques result in less-expensive and less time consuming methods of semiconductor fabrication.

Production

Quantum Dots with gradually stepping emission from violet to deep red are being produced in a kg scale at PlasmaChem GmbH

There are several ways to prepare quantum dots, the principal ones involving colloids.

Colloidal Synthesis

Colloidal semiconductor nanocrystals are synthesized from solutions, much like traditional chemical processes. The main difference is the product neither precipitates as a bulk solid nor remains dissolved. Heating the solution at high temperature, the precursors decompose forming monomers which then nucleate and generate nanocrystals. Temperature is a critical factor in determining optimal conditions for the nanocrystal growth. It must be high enough to allow for rearrangement and annealing of atoms during the synthesis process while being low enough to promote crystal growth. The concentration of monomers is another critical factor that has to be stringently controlled during nanocrystal growth. The growth process of nanocrystals can occur in two different regimes, "focusing" and "defocusing". At high monomer concentrations, the critical size (the size where nanocrystals neither grow nor shrink) is relatively small, resulting in growth of nearly all particles. In this regime, smaller particles grow faster than large ones (since larger crystals need more atoms to grow than small crystals) resulting in "focusing" of the size distribution to yield nearly monodisperse particles. The size focusing is optimal when the monomer concentration is kept such that the average nanocrystal size present is always slightly larger than the critical size. Over time, the monomer concentration diminishes, the critical size becomes larger than the average size present, and the distribution "defocuses".

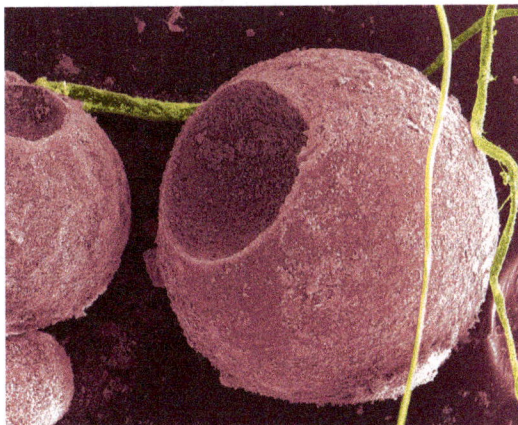

Cadmium sulfide quantum dots on cells

There are colloidal methods to produce many different semiconductors. Typical dots are made of binary compounds such as lead sulfide, lead selenide, cadmium selenide, cadmium sulfide, indium arsenide, and indium phosphide. Dots may also be made from ternary compounds such as cadmium selenide sulfide. These quantum dots can contain as few as 100 to 100,000 atoms within the quantum dot volume, with a diameter of ~ 10 to 50 atoms. This corresponds to about 2 to 10 nanometers, and at 10 nm in diameter, nearly 3 million quantum dots could be lined up end to end and fit within the width of a human thumb.

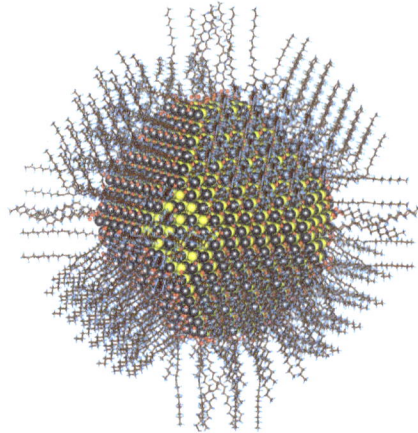

Ideallized image of colloidal nanoparticle of lead sulfide (selenide) with complete passivation by oleic acid, oleyl amine and hydroxyl ligands (size ~5nm)

Large batches of quantum dots may be synthesized via colloidal synthesis. Due to this scalability and the convenience of benchtop conditions, colloidal synthetic methods are promising for commercial applications. It is acknowledged to be the least toxic of all the different forms of synthesis.

Plasma Synthesis

Plasma synthesis has evolved to be one of the most popular gas-phase approaches for the production of quantum dots, especially those with covalent bonds. For example, silicon (Si) and germanium (Ge) quantum dots have been synthesized by using nonthermal plasma. The size, shape, surface and composition of quantum dots can all be controlled in nonthermal plasma. Doping that seems quite challenging for quantum dots has also been realized in plasma synthesis. Quantum dots synthesized by plasma are usually in the form of powder, for which surface modification may be carried out. This can lead to excellent dispersion of quantum dots in either organic solvents or water (i. e., colloidal quantum dots).

Fabrication

- Self-assembled quantum dots are typically between 5 and 50 nm in size. Quantum dots defined by lithographically patterned gate electrodes, or by etching on two-dimensional electron gasses in semiconductor heterostructures can have lateral dimensions between 20 and 100 nm.

- Some quantum dots are small regions of one material buried in another with a larger band gap. These can be so-called core–shell structures, e.g., with CdSe in the core and ZnS in the shell or from special forms of silica called ormosil.

- Quantum dots sometimes occur spontaneously in quantum well structures due to monolayer fluctuations in the well's thickness.

- Self-assembled quantum dots nucleate spontaneously under certain conditions during molecular beam epitaxy (MBE) and metallorganic vapor phase epitaxy (MOVPE), when a material is grown on a substrate to which it is not lattice matched. The resulting strain produces coherently strained islands on top of a two-dimensional wetting layer. This growth mode is known as Stranski–Krastanov growth. The islands can be subsequently buried to form the quantum dot. This fabrication method has potential for applications in quantum cryptography (i.e. single photon sources) and quantum computation. The main limitations of this method are the cost of fabrication and the lack of control over positioning of individual dots.

- Individual quantum dots can be created from two-dimensional electron or hole gases present in remotely doped quantum wells or semiconductor heterostructures called lateral quantum dots. The sample surface is coated with a thin layer of resist. A lateral pattern is then defined in the resist by electron beam lithography. This pattern can then be transferred to the electron or hole gas by etching, or by depositing metal electrodes (lift-off process) that allow the application of external voltages between the electron gas and the electrodes. Such quantum dots are mainly of interest for experiments and applications involving electron or hole transport, i.e., an electrical current.

- The energy spectrum of a quantum dot can be engineered by controlling the geometrical size, shape, and the strength of the confinement potential. Also, in contrast to atoms, it is relatively easy to connect quantum dots by tunnel barriers to conducting leads, which allows the application of the techniques of tunneling spectroscopy for their investigation.

The quantum dot absorption features correspond to transitions between discrete, three-dimensional particle in a box states of the electron and the hole, both confined to the same nanometer-size box. These discrete transitions are reminiscent of atomic spectra and have resulted in quantum dots also being called *artificial atoms*.

- Confinement in quantum dots can also arise from electrostatic potentials (generated by external electrodes, doping, strain, or impurities).

- Complementary metal-oxide-semiconductor (CMOS) technology can be employed to fabricate silicon quantum dots. Ultra small (L=20 nm, W=20 nm) CMOS transistors behave as single electron quantum dots when operated at cryogenic temperature over a range of −269 °C (4 K) to about −258 °C (15 K). The transistor displays Coulomb blockade due to progressive charging of electrons one by one. The number of electrons confined in the channel is driven by the gate voltage, starting from an occupation of zero electrons, and it can be set to 1 or many.

Viral Assembly

Genetically engineered M13 bacteriophage viruses allow preparation of quantum dot biocomposite structures. It had previously been shown that genetically engineered viruses can recognize specific semiconductor surfaces through the method of selection by combinatorial phage display. Additionally, it is known that liquid crystalline structures of wild-type viruses (Fd, M13, and TMV) are adjustable by controlling the solution concentrations, solution ionic strength, and the external magnetic field applied to the solutions. Consequently, the specific recognition properties of the virus can be used to organize inorganic nanocrystals, forming ordered arrays over the length scale defined by liquid crystal formation. Using this information, Lee et al. (2000) were able to create self-assembled, highly oriented, self-supporting films from a phage and ZnS precursor solution. This system allowed them to vary both the length of bacteriophage and the type of inorganic material through genetic modification and selection.

Electrochemical Assembly

Highly ordered arrays of quantum dots may also be self-assembled by electrochemical techniques. A template is created by causing an ionic reaction at an electrolyte-metal interface which results in the spontaneous assembly of nanostructures, including quantum dots, onto the metal which is then used as a mask for mesa-etching these nanostructures on a chosen substrate.

Bulk-manufacture

Quantum dot manufacturing relies on a process called "high temperature dual injection" which has been scaled by multiple companies for commercial applications that require large quantities (hundreds of kilograms to tonnes) of quantum dots. This reproducible production method can be applied to a wide range of quantum dot sizes and compositions.

The bonding in certain cadmium-free quantum dots, such as III-V-based quantum dots, is more covalent than that in II-VI materials, therefore it is more difficult to separate nanoparticle nucleation and growth via a high temperature dual injection synthesis. An alternative method of quantum dot synthesis, the "molecular seeding" process, provides a reproducible route to the production of high quality quantum dots in large volumes. The process utilises identical molecules of a molecular cluster compound as the nucleation sites for nanoparticle growth, thus avoiding the need for a high temperature injection step. Particle growth is maintained by the periodic addition of precursors at moderate temperatures until the desired particle size is reached. The molecular seeding process is not limited to the production of cadmium-free quantum dots; for example, the process can be used to synthesise kilogram batches of high quality II-VI quantum dots in just a few hours.

Another approach for the mass production of colloidal quantum dots can be seen in the transfer of the well-known hot-injection methodology for the synthesis to a technical continuous flow system. The batch-to-batch variations arising from the needs during the mentioned methodology can be overcome by utilizing technical components for mixing and growth as well as transport and temperature adjustments. For the production of CdSe based semiconductor nanoparticles this method has been investigated and tuned to production amounts of kg per month. Since the use of technical components allows for easy interchange in regards of maximum through-put and size, it can be further enhanced to tens or even hundreds of kilograms.

In 2011 a consortium of U.S. and Dutch companies reported a "milestone" in high volume quantum dot manufacturing by applying the traditional high temperature dual injection method to a flow system.

On January 23, 2013 Dow entered into an exclusive licensing agreement with UK-based Nanoco for the use of their low-temperature molecular seeding method for bulk manufacture of cadmium-free quantum dots for electronic displays, and on September 24, 2014 Dow commenced work on the production facility in South Korea capable of producing sufficient quantum dots for "millions of cadmium-free televisions and other devices, such as tablets". Mass production is due to commence in mid-2015. On 24 March 2015 Dow announced a partnership deal with LG Electronics to develop the use of cadmium free quantum dots in displays.

Heavy Metal-free Quantum Dots

In many regions of the world there is now a restriction or ban on the use of heavy metals in many household goods, which means that most cadmium based quantum dots are unusable for consumer-goods applications.

For commercial viability, a range of restricted, heavy metal-free quantum dots has been developed showing bright emissions in the visible and near infra-red region of the spectrum and have similar optical properties to those of CdSe quantum dots. Among these systems are InP/ZnS and CuInS/ZnS, for example.

Peptides are being researched as potential quantum dot material. Since peptides occur naturally in all organisms, such dots would likely be nontoxic and easily biodegraded.

Safety

Toxicity

Some quantum dots pose risks to human health and the environment under certain conditions. Notably, the studies on quantum dot toxicity are focused on cadmium containing particles and has yet to be demonstrated in animal models after physiologically relevant dosing. In vitro studies, based on cell cultures, on quantum dots (QD) toxicity suggests that their toxicity may derive from multiple factors including its physicochemical characteristics (size, shape, composition, surface functional groups, and surface charges) and environment. Assessing their potential toxicity is complex as these factors include properties such as QD size, charge, concentration, chemical composition, capping ligands, and also on their oxidative, mechanical and photolytic stability.

Many studies have focused on the mechanism of QD cytotoxicity using model cell cultures. It has been demonstrated that after exposure to ultraviolet radiation or oxidized by air CdSe QDs release free cadmium ions causing cell death. Group II-VI QDs also have been reported to induce the formation of reactive oxygen species after exposure to light, which in turn can damage cellular components such as proteins, lipids and DNA. Some studies have also demonstrated that addition of a ZnS shell inhibit the process of reactive oxygen species in CdSe QDs. Another aspect of QD toxicity is the process of their size dependent intracellular pathways that concentrate these particles in cellular organelles that are inaccessible by metal ions, which may result in unique patterns of cytotoxicity compared to their constituent metal ions. The reports of QD localization in the cell nucleus present additional modes of toxicity because they may induce DNA mutation, which in turn will propagate through future generation of cells causing diseases.

Although concentration of QDs in certain organelles have been reported in in vivo studies using animal models, interestingly, no alterations in animal behavior, weight, hematological markers or organ damage has been found through either histological or biochemical analysis. These finding have led scientists to believe that intracellular dose is the most important deterring factor for QD toxicity. Therefore, factors determining the QD endocytosis that determine the effective intracellular concentration, such as QD size, shape and surface chemistry determine their toxicity. Excretion of QDs through urine in animal models also have demonstrated via injecting radio-labeled ZnS capped CdSe QDs where the ligand shell was labelled with [99mTc]. Though multiple other studies have concluded retention of QDs in cellular levels, exocytosis of QDs is still poorly studied in the literature.

While significant research efforts have broadened the understanding of toxicity of QDs, there are large discrepancies in the literature and questions still remains to be answered. Diversity of this class material as compared to normal chemical substances makes the assessment of their toxicity very challenging. As their toxicity may also be dynamic depending on the environmental factors such as pH level, light exposure and cell type, traditional methods of assessing toxicity of chemicals such as LD_{50} are not applicable for QDs. Therefore, researchers are focusing on introducing novel approaches and adapting existing methods to include this unique class of materials. Furthermore, novel strategies to engineer safer QDs are still under exploration by the scientific community. A recent novelty in the field is the discovery of carbon quantum dots, a new generation of optically-active nanoparticles potentially capable of replacing semiconductor QDs, but with the advantage of much lower toxicity.

Optical Properties

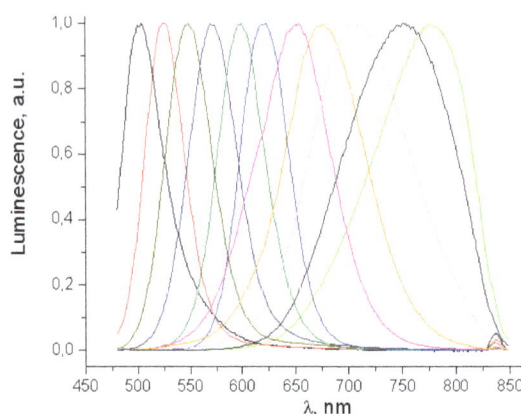

Fluorescence spectra of CdTe quantum dots of various sizes. Different sized quantum dots emit different color light due to quantum confinement.

In semiconductors, light absorption generally leads to an electron being excited from the valence to the conduction band, leaving behind a hole. The electron and the hole can bind to each other to form an exciton. When this exciton recombines (i.e. the electron resumes its ground state), the exciton's energy can be emitted as light. This is called fluorescence. In a simplified model, the energy of the emitted photon can be understood as the sum of the band gap energy between the highest occupied level and the lowest unoccupied energy level, the confinement energies of the hole and the excited electron, and the bound energy of the exciton (the electron-hole pair):

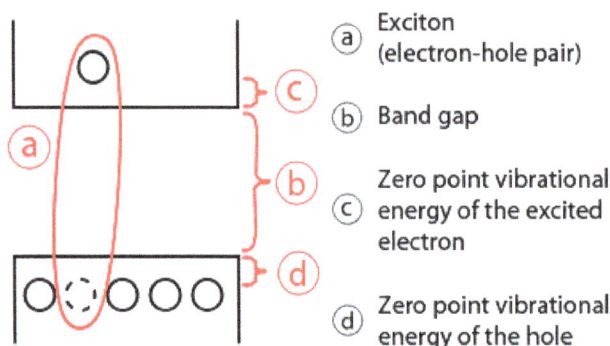

As the confinement energy depends on the quantum dot's size, both absorption onset and fluorescence emission can be tuned by changing the size of the quantum dot during its synthesis.

The larger the dot, the redder (lower energy) its absorption onset and fluorescence spectrum. Conversely, smaller dots absorb and emit bluer (higher energy) light. Recent articles in *Nanotechnology* and in other journals have begun to suggest that the shape of the quantum dot may be a factor in the coloration as well, but as yet not enough information is available. Furthermore, it was shown that the lifetime of fluorescence is determined by the size of the quantum dot. Larger dots have more closely spaced energy levels in which the electron-hole pair can be trapped. Therefore, electron-hole pairs in larger dots live longer causing larger dots to show a longer lifetime.

To improve fluorescence quantum yield, quantum dots can be made with "shells" of a larger bandgap semiconductor material around them. The improvement is suggested to be due to the reduced access of electron and hole to non-radiative surface recombination pathways in some cases, but also due to reduced auger recombination in others.

Potential Applications

Quantum dots are particularly promising for optical applications due to their high extinction coefficient. They operate like a single electron transistor and show the Coulomb blockade effect. Quantum dots have also been suggested as implementations of qubits for quantum information processing.

Tuning the size of quantum dots is attractive for many potential applications. For instance, larger quantum dots have a greater spectrum-shift towards red compared to smaller dots, and exhibit less pronounced quantum properties. Conversely, the smaller particles allow one to take advantage of more subtle quantum effects.

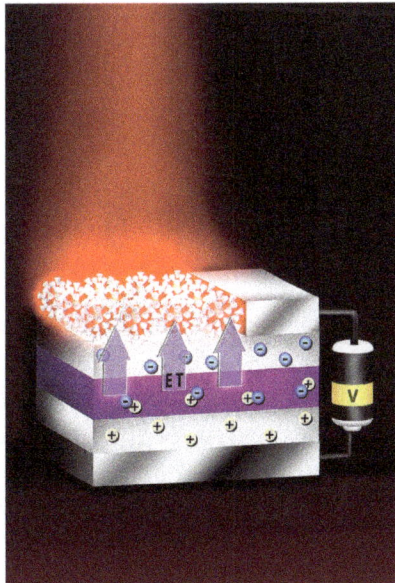

A device that produces visible light, through energy transfer from thin layers of quantum wells to crystals above the layers.

Being zero-dimensional, quantum dots have a sharper density of states than higher-dimensional structures. As a result, they have superior transport and optical properties. They have potential uses in diode lasers, amplifiers, and biological sensors. Quantum dots may be excited within a locally enhanced electromagnetic field produced by gold nanoparticles, which can then be observed from the surface plasmon resonance in the photoluminescent excitation spectrum of (CdSe)ZnS nanocrystals. High-quality quantum dots are well suited for optical encoding and multiplexing applications due to their broad excitation profiles and narrow/symmetric emission spectra. The new generations of quantum dots have far-reaching potential for the study of intracellular processes at the single-molecule level, high-resolution cellular imaging, long-term in vivo observation of cell trafficking, tumor targeting, and diagnostics.

CdSe nanocrystals are efficient triplet photosensitizers. Laser excitation of small CdSe nanoparticles enables the extraction of the excited state energy from the Quantum Dots into bulk solution, thus opening the door to a wide range of potential applications such as photodynamic therapy, photovoltaic devices, molecular electronics, and catalysis.

Computing

Quantum dot technology is potentially relevant to solid-state quantum computation. By applying small voltages to the leads, current through the quantum dot can be controlled and thereby precise measurements of the spin and other properties therein can be made. With several entangled quantum dots, or qubits, plus a way of performing operations, quantum calculations and the computers that would perform them might be possible.

Biology

In modern biological analysis, various kinds of organic dyes are used. However, as technology advances, greater flexibility in these dyes is sought. To this end, quantum dots have quickly filled in the role, being found to be superior to traditional organic dyes on several counts, one of the most immediately obvious being brightness (owing to the high extinction coefficient combined with a comparable quantum yield to fluorescent dyes) as well as their stability (allowing much less photobleaching). It has been estimated that quantum dots are 20 times brighter and 100 times more stable than traditional fluorescent reporters. For single-particle tracking, the irregular blinking of quantum dots is a minor drawback. However, there have been groups which have developed quantum dots which are essentially nonblinking and demonstrated their utility in single molecule tracking experiments.

The use of quantum dots for highly sensitive cellular imaging has seen major advances. The improved photostability of quantum dots, for example, allows the acquisition of many consecutive focal-plane images that can be reconstructed into a high-resolution three-dimensional image. Another application that takes advantage of the extraordinary photostability of quantum dot probes is the real-time tracking of molecules and cells over extended periods of time. Antibodies, streptavidin, peptides, DNA, nucleic acid aptamers, or small-molecule ligands can be used to target quantum dots to specific proteins on cells. Researchers were able to observe quantum dots in lymph nodes of mice for more than 4 months.

Semiconductor quantum dots have also been employed for in vitro imaging of pre-labeled cells. The ability to image single-cell migration in real time is expected to be important to several research areas such as embryogenesis, cancer metastasis, stem cell therapeutics, and lymphocyte immunology.

One application of quantum dots in biology is as donor fluorophores in Förster resonance energy transfer, where the large extinction coefficient and spectral purity of these fluorophores make them superior to molecular fluorophores It is also worth noting that the broad absorbance of QDs allows selective excitation of the QD donor and a minimum excitation of a dye acceptor in FRET-based studies. The applicability of the FRET model, which assumes that the Quantum Dot can be approximated as a point dipole, has recently been demonstrated

The use of quantum dots for tumor targeting under in vivo conditions employ two targeting schemes: active targeting and passive targeting. In the case of active targeting, quantum dots are functionalized with tumor-specific binding sites to selectively bind to tumor cells. Passive targeting uses the enhanced permeation and retention of tumor cells for the delivery of quantum dot probes. Fast-growing tumor cells typically have more permeable membranes than healthy cells, allowing the leakage of small nanoparticles into the cell body. Moreover, tumor cells lack an effective lymphatic drainage system, which leads to subsequent nanoparticle-accumulation.

Quantum dot probes exhibit in vivo toxicity. For example, CdSe nanocrystals are highly toxic to cultured cells under UV illumination, because the particles dissolve, in a process known as photolysis, to release toxic cadmium ions into the culture medium. In the absence of UV irradiation, however, quantum dots with a stable polymer coating have been found to be essentially nontoxic. Hydrogel encapsulation of quantum dots allows for quantum dots to be introduced into a stable aqueous solution, reducing the possibility of cadmium leakage. Then again, only little is known about the excretion process of quantum dots from living organisms.

In another potential application, quantum dots are being investigated as the inorganic fluorophore for intra-operative detection of tumors using fluorescence spectroscopy.

Delivery of undamaged quantum dots to the cell cytoplasm has been a challenge with existing techniques. Vector-based methods have resulted in aggregation and endosomal sequestration of quantum dots while electroporation can damage the semi-conducting particles and aggregate delivered dots in the cytosol. Via cell squeezing, quantum dots can be efficiently delivered without inducing aggregation, trapping material in endosomes, or significant loss of cell viability. Moreover, it has shown that individual quantum dots delivered by this approach are detectable in the cell cytosol, thus illustrating the potential of this technique for single molecule tracking studies.

Photovoltaic Devices

The tunable absorption spectrum and high extinction coefficients of quantum dots make them attractive for light harvesting technologies such as photovoltaics. Quantum dots may be able to increase the efficiency and reduce the cost of today's typical silicon photovoltaic cells. According to an experimental proof from 2004, quantum dots of lead selenide can produce more than one exciton from one high energy photon via the process of carrier multiplication or multiple exciton generation (MEG). This compares favorably to today's photovoltaic cells which can only manage one exciton per high-energy photon, with high kinetic energy carriers losing their energy as heat. Quantum dot photovoltaics would theoretically be cheaper to manufacture, as they can be made "using simple chemical reactions."

Quantum Dot only Solar Cells

Aromatic self-assembled monolayers (SAMs) (e.g. 4-nitrobenzoic acid) can be used to improve the band alignment at electrodes for better efficiencies. This technique has provided a record power conversion efficiency (PCE) of 10.7%. The SAM is positioned between ZnO-PbS colloidal quantum dot (CQD) film junction to modify band alignment via the dipole moment of the constituent SAM molecule, and the band tuning may be modified via the density, dipole and the orientation of the SAM molecule.

Quantum Dot in Hybrid Solar Cells

Colloidal quantum dots are also used in inorganic/organic hybrid solar cells. These solar cells are attractive because of potentially their low-cost fabrication and relatively high efficiency. Incorporation of metal oxides, such as ZnO, TiO_2, and Nb_2O_5 nanomaterials into organic photovoltaics have been commercialized using full roll-to-roll processing. A 13.2% power conversion efficiency is claimed in Si nanowire/PEDOT:PSS hybrid solar cells.

Quantum Dot with Nanowire in Solar Cells

Another potential use involves capped single-crystal ZnO nanowires with CdSe quantum dots, immersed in mercaptopropionic acid as hole transport medium in order to obtain a QD-sensitized solar cell. The morphology of the nanowires allowed the electrons to have a direct pathway to the photoanode. This form of solar cell exhibits 50-60% internal quantum efficiencies.

Nanowires with quantum dot coatings on silicon nanowires (SiNW) and carbon quantum dots. The use of SiNWs instead of planar silicon enhances the antiflection properties of Si. The SiNW exhibits a light-trapping effect due to light trapping in the SiNW. This use of SiNWs in conjunction with carbon quantum dots resulted in a solar cell that reached 9.10% PCE.

Graphene quantum dots have also been blended with organic electronic materials to improve efficiency and lower cost in photovoltaic devices and organic light emitting diodes (OLEDs) in compared to graphene sheets. These graphene quantum dots were functionalized with organic ligands that experience photoluminescence from UV-Vis absorption.

Light Emitting Devices

Several methods are proposed for using quantum dots to improve existing light-emitting diode (LED) design, including "Quantum Dot Light Emitting Diode" (QD-LED) displays and "Quantum Dot White Light Emitting Diode" (QD-WLED) displays. Because Quantum dots naturally produce monochromatic light, they can be more efficient than light sources which must be color filtered. QD-LEDs can be fabricated on a silicon substrate, which allows them to be integrated onto standard silicon-based integrated circuits or microelectromechanical systems. Quantum dots are valued for displays, because they emit light in very specific gaussian distributions. This can result in a display with visibly more accurate colors. A conventional color liquid crystal display (LCD) is usually backlit by fluorescent lamps (CCFLs) or conventional white LEDs that are color filtered to produce red, green, and blue pixels. An improvement is using conventional blue-emitting LEDs as the light sources and converting part of the emitted light into *pure* green and red light by the appropriate quantum dots placed in front of the blue LED or using a quantum dot infused diffuser sheet in the backlight optical stack. This type of white light as the backlight of an LCD panel allows for the best color gamut at lower cost than a RGB LED combination using three LEDs.

The ability of QDs to precisely convert and tune a spectrum makes them attractive for LCD displays. Previous LCD displays can waste energy converting red-green poor, blue-yellow rich white light into a more balanced lighting. By using QDs, only the necessary colors for ideal images are contained in the screen. The result is a screen that is brighter, clearer, and more energy-efficient. The first commercial application of quantum dots was the Sony XBR X900A series of flat panel televisions released in 2013.

In June 2006, QD Vision announced technical success in making a proof-of-concept quantum dot display and show a bright emission in the visible and near infra-red region of the spectrum. A QD-LED integrated at a scanning microscopy tip was used to demonstrate fluorescence near-field scanning optical microscopy (NSOM) imaging.

Photodetector Devices

Quantum dot photodetectors (QDPs) can be fabricated either via solution-processing, or from conventional single-crystalline semiconductors. Conventional single-crystalline semiconductor QDPs are precluded from integration with flexible organic electronics due to the incompatibility of their growth conditions with the process windows required by organic semiconductors. On the other hand, solution-processed QDPs can be readily integrated with an almost infinite variety of substrates, and also postprocessed atop other integrated circuits. Such colloidal QDPs have potential applications in surveillance, machine vision, industrial inspection, spectroscopy, and fluorescent biomedical imaging.

Photocatalysts

Quantum dots also function as photocatalysts for the light driven chemical conversion of water into hydrogen as a pathway to solar fuel. In photocatalysis, electron hole pairs formed in the dot under band gap excitation drive redox reactions in the surrounding liquid. Generally, the photocatalytic activity of the dots is related to the particle size and its degree of quantum confinement. This is because the band gap determines the chemical energy that is stored in the dot in the excited state. An obstacle for the use of quantum dots in photocatalysis is the presence of surfactants on the surface of the dots. These surfactants (or ligands) interfere with the chemical reactivity of the dots by slowing down mass transfer and electron transfer processes. Also, quantum dots made of metal chalcogenides are chemically unstable under oxidizing conditions and undergo photo corrosion reactions.

Theory

Quantum dots are theoretically described as a point like, or a zero dimensional (0D) entity. Most of their properties depend on the dimensions, shape and materials of which QDs are made. Generally QDs present different thermodynamic properties from the bulk materials of which they are made. One of these effects is the Melting-point depression. Optical properties of spherical metallic QDs are well described by the Mie scattering theory.

Quantum Confinement in Semiconductors

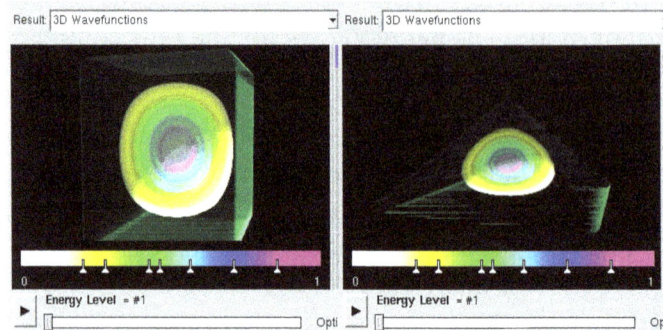

3D confined electron wave functions in a quantum dot. Here, rectangular and triangular-shaped quantum dots are shown. Energy states in rectangular dots are more *s-type* and *p-type*. However, in a triangular dot the wave functions are mixed due to confinement symmetry.

In a semiconductor crystallite whose size is smaller than twice the size of its exciton Bohr radius, the excitons are squeezed, leading to quantum confinement. The energy levels can then be predicted using the particle in a box model in which the energies of states depends on the length of the box. Comparing the quantum dots size to the Bohr radius of the electron and hole wave functions, 3 regimes can be defined. A 'strong confinement regime' is defined as the quantum dots radius being smaller than both electron and hole Bohr radius, 'weak confinement' is given when the quantum dot is larger than both. For semiconductors in which electron and hole radii are markedly different, an 'intermediate confinement regime' exists, where the quantum dot's radius is larger than the Bohr radius of one charge carrier (typically the hole), but not the other charge carrier.

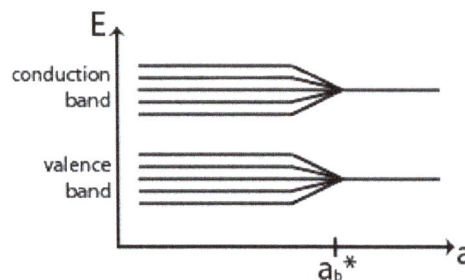

Splitting of energy levels for small quantum dots due to the quantum confinement effect. The horizontal axis is the radius, or the size, of the quantum dots and a_b^* is the Exciton Bohr radius.

Band Gap Energy

The band gap can become smaller in the strong confinement regime as the energy levels split up. The Exciton Bohr radius can be expressed as:

$$a_b^* = \varepsilon_r \left(\frac{m}{\mu} \right) a_b$$

where a_b is the Bohr radius=0.053 nm, m is the mass, μ is the reduced mass, and ε_r is the size-dependent dielectric constant (Relative permittivity). This results in the increase in the total emission energy (the sum of the energy levels in the smaller band gaps in the strong confinement regime is larger than the energy levels in the band gaps of the original levels in the weak confinement regime) and the emission at various wavelengths. If the size distribution of QDs is not enough peaked, the convolution of multiple emission wavelengths is observed as a continuous spectra.

Confinement Energy

The exciton entity can be modeled using the particle in the box. The electron and the hole can be seen as hydrogen in the Bohr model with the hydrogen nucleus replaced by the hole of positive charge and negative electron mass. Then the energy levels of the exciton can be represented as the solution to the particle in a box at the ground level (n = 1) with the mass replaced by the reduced mass. Thus by varying the size of the quantum dot, the confinement energy of the exciton can be controlled.

Bound Exciton Energy

There is Coulomb attraction between the negatively charged electron and the positively charged hole. The negative energy involved in the attraction is proportional to Rydberg's energy and inversely proportional to square of the size-dependent dielectric constant of the semiconductor. When the size of the semiconductor crystal is smaller than the Exciton Bohr radius, the Coulomb interaction must be modified to fit the situation.

Therefore, the sum of these energies can be represented as:

$$E_{confinement} = \frac{\hbar^2 \pi^2}{2a^2}\left(\frac{1}{m_e} + \frac{1}{m_h}\right) = \frac{\hbar^2 \pi^2}{2\mu a^2}$$

$$E_{exciton} = -\frac{1}{\epsilon_r^2}\frac{\mu}{m_e}R_y = -R_y^*$$

$$E = E_{bandgap} + E_{confinement} + E_{exciton}$$

$$= E_{bandgap} + \frac{\hbar^2 \pi^2}{2\mu a^2} - R_y^*$$

where μ is the reduced mass, a is the radius, m_e is the free electron mass, m_h is the hole mass, and ε_r is the size-dependent dielectric constant.

Although the above equations were derived using simplifying assumptions, they imply that the electronic transitions of the quantum dots will depend on their size. These quantum confinement effects are apparent only below the critical size. Larger particles do not exhibit this effect. This effect of quantum confinement on the quantum dots has been repeatedly verified experimentally and is a key feature of many emerging electronic structures.

The Coulomb interaction between confined carriers can also be studied by numerical means when results unconstrained by asymptotic approximations are pursued.

Besides confinement in all three dimensions (i.e., a quantum dot), other quantum confined semiconductors include:

- Quantum wires, which confine electrons or holes in two spatial dimensions and allow free propagation in the third.

- Quantum wells, which confine electrons or holes in one dimension and allow free propagation in two dimensions.

Models

A variety of theoretical frameworks exist to model optical, electronic, and structural properties of quantum dots. These may be broadly divided into quantum mechanical, semiclassical, and classical.

Quantum Mechanics

Quantum mechanical models and simulations of quantum dots often involve the interaction of electrons with a pseudopotential or random matrix.

Semiclassical

Semiclassical models of quantum dots frequently incorporate a chemical potential. For example, The thermodynamic chemical potential of an N-particle system is given by

$$\mu(N) = E(N) - E(N-1)$$

whose energy terms may be obtained as solutions of the Schrödinger equation. The definition of capacitance,

$$\frac{1}{C} \equiv \frac{\Delta V}{\Delta Q},$$

with the potential difference

$$V = \frac{\Delta\mu}{e} = \frac{\mu(N+\Delta N) - \mu(N)}{e}$$

may be applied to a quantum dot with the addition or removal of individual electrons,

$$\Delta N = 1 \text{ and } \Delta Q = e.$$

Then

$$C(N) = \frac{e^2}{\mu(N+1) - \mu(N)} = \frac{e^2}{I(N) - A(N)}$$

is the "quantum capacitance" of a quantum dot, where we denoted by $I(N)$ the ionization potential and by $A(N)$ the electron affinity of the N-particle system.

Classical Mechanics

Classical models of electrostatic properties of electrons in quantum dots are similar in nature to the Thomson problem of optimally distributing electrons on a unit sphere.

The classical electrostatic treatment of electrons confined to spherical quantum dots is similar to their treatment in the Thomson, or plum pudding model, of the atom.

The classical treatment of both two-dimensional and three-dimensional quantum dots exhibit electron shell-filling behavior. A "periodic table of classical artificial atoms" has been described for two-dimensional quantum dots. As well, several connections have been reported between the three-dimensional Thomson problem and electron shell-filling patterns found in naturally-occurring atoms found throughout the periodic table. This latter work originated in classical electrostatic modeling of electrons in a spherical quantum dot represented by an ideal dielectric sphere.

History

The term "quantum dot" was coined in 1988. They were first discovered in a glass matrix and in colloidal solutions by Alexey Ekimov.

Gold Nanoparticles as Biosensors

The combination of nanotechnology with chemistry, biology, physics, and medicine for the development of ultrasensitive detection and imaging methods in analytical or biological sciences is becoming increasingly important in modern day science [1–14]. Particularly attractive is the use of functional GNPs in biological and pharmaceutical fi eld, such as the ultrasensitive detection and imaging methods for bio reorganizing events, because GNPs have unique optical properties (i.e. surface plasma resonance absorption and resonance light scattering), a variety of surface coatings and great biocompatibility. Because nanoparticles have a high surface area to volume ratio, the plasmon frequency is exquisitely sensitive to the dielectric (refractive index) nature of its interface with the local medium. Any changes to the environment of these particles (surface modification, aggregation, medium refractive index, etc.) may lead to colorimetric changes of the dispersions.

GNPs are useful in a broad range of applications, but practical limitations are apparent when mono dispersity is required. Numerous preparative methods for GNPs from about 1 nm to several micrometers diameter have been documented in literature [15-26]. The most widely applied procedures to obtain 10 to 150 nm GNPs are variations of the classic Turkevich– Frenscitrate reduction of gold (III) derivatives [15-16]. GNPs size (between 10 and 147 nm) can be controlled by the ratio between the reducing/stabilizing agents (trisodium citrate) and gold (III) derivatives (hydrogen/sodium tetrachloroaurate (III)). This method is used frequently these days as the lose shell of citrate on the particle-surfaces can be easily replaced by other desired ligands (e.g., thiolated DNA). Most hydrophobic GNPs (also sometimes called monolayer protected clusters (MPCs)) with diameters in 1 to 8 nm ranges are prepared by the Brust– Schiffrin method: the gold (III) derivatives are reduced by sodium borohydride (NaBH4) in an organic solvent in the presence of thiol capping ligands using either a two-phase liquid/liquid system or a suitable single-phase solvent [18,19]. In the Brust–Schiffrin methods, tetrachloroaurate (III) is transferred to toluene using tetraoctylammoniumbromide (TOAB) as the phase-transfer reagent and reduced by NaBH4 in the presence of dodecanethiol (DDT). Larger thiol/gold mole ratios give smaller average core sizes, and fast reductant addition and cooled solutions produced smaller, more monodisperse particles.

For many applications, attaching functional groups of interest to the nanoparticles has to be readily achieved taking into consideration that the probes must not bind non-specifically to each other or to anything else present in the system under investigation. In addition, introducing multiple functionalities could be of great value, as it provides more fl exibility for multiplexing in bioanalytical applications. Electrostatic interaction, specific recognition (antibody–antigen, biotin–avidin, etc.), and covalent coupling (Au–S covalent, etc.) are three kinds of widely used methods to synthesis GNP probes to meet the application requirement.

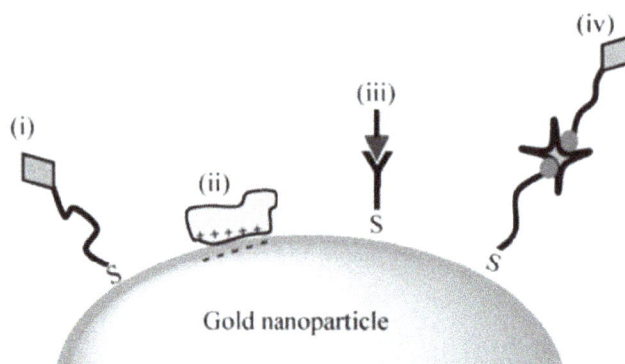

(i): Thiolated or disulfide modified ligands
(ii): Electrostatic interaction
(iii): Antibody-antigen associations
(iv): Streptavidin-biotin binding

Synthesis of gold nanoparticle probes

Electrostatic interaction or physical adsorption of ligands for GNP probes is a simple process with the benefits of time saving and reduced complexity of ligand preparation [6, 27-28].

Its relative simplicity gives this approach certain advantages over the more complex covalent immobilization methods. However, the binding is not strong enough to yield stable surfaces capable of standing the necessary washing steps and incubation conditions in biological studies on subsequent reaction. This issue is even more crucial in biological studies under harsh experimental conditions, for example, long time incubation with buffer solution containing attacking molecules such as dithiothreitol (a small, uncharged molecule with two thiol groups, used to protect proteins from oxidation) as well as high salt concentration (generally used in DNA hybridization experiment). This results in a strong non-specific interaction between the GNP probes and analytes which leads to decreased detection selectivity.

Whereas covalent binding is normally more complex, sometimes requiring intensive synthesis works on the ligands. On the other hand, covalent binding of ligands with GNPs offers high stability and is demonstrated to be quite robust as the probes so prepared can withstand a very high salt concentration (e.g., 2M NaCl) and are extremely stable at high temperatures.

Applications of Gold Nanoparticles as Sensor

Potential applications of GNPs as sensor in analytical and/or biological sciences include chemical sensing and imaging applications. The color changes associated with nanoparticle aggregation were originally exploited by Mirkin et. al. who showed that ssDNA stabilized nanoparticles could be used to colorimetrically detect the complementary oligonucleotide [29- 30]. The

Mirkin's type colorimetrically assay has opened up new avenues to apply GNPs. In the past few years, numerous GNP based assays have been developed for the detection of many targets, including: metal ions, small organic compounds, nucleic acids, proteins, cells, etc.

Heavy Metal Cations Determination

Heavy metal cations such as Pb^{2+}, Cr^{2+} and Hg^{2+} are the commonly encountered toxic substances in the environment and pose significant public health hazards when they are present in drinking water even in parts per million concentrations [33-35]. In particular, Pb^{2+} is dangerous for children, causing mental retardation . Colorimetric sensors using GNPs have been widely explored and have important applications in the sensitive detection of metallic ions [36-47]. The GNP-based colorimetric assay does not utilize organic solvents, enzymatic reactions, light-sensitive dye molecules, lengthy protocols, or sophisticated instrumentation thereby over coming some of the limitations of the conventional methods. Lu's group has developed a fast and simple colorimetric sensor for on-site and real-time heavy metal cation detection based on a DNAzyme modification of GNP [43-48]. The sensor has a detection limit of 3 nM for Pb^{2+}, which is much lower than the Environmental Protection Agency of United States (EPA) limit for lead ions in drinking water [43-46].

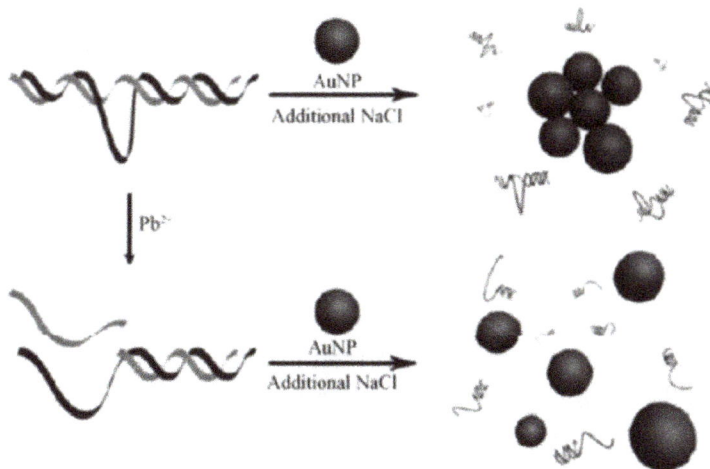

GNPs as sensor to detect PB^{2+}

Willner and co-workers developed a method for colorimetric detection of Hg^{2+} ions by a DNA based machine . This method reveals a substantial enhancement in the sensitivity (1 nm, 0.2 ppb) over the reported methods, and comparable sensitivity to the reported DNAzyme method. Based on the Hg^{2+} mediated formation of T-Hg^{2+}-T base pairs, a highly sensitive and selective Hg^{2+} detection assay has also been developed by Liu's group and Mirkin's group [49- 50]. This is based on Hg^{2+} induced thymine–thymine (T–T) mismatches in DNA modified GNPs to form particle aggregates at room temperature with a concomitant colorimetric response. This method is enzyme free and does not require specialized equipment other than a temperature control unit. The concentration of Hg^{2+} can be determined by the change of the solution color at a given temperature or the melting temperature (T_m) of the DNA–GNP aggregates. Significantly, this method can, in principle, be used to detect other metal ions by substituting the thymidine in study with synthetic artificial bases that selectively bind other metal ions.

Detection of Small Molecules

The GNP based colorimetric methods have also been used for detecting small molecules. In these assays, GNPs were functionalized with groups that provide affinity sites for the binding of small molecules . Lu and coworkers took advantage of the color change of GNPs induced by hydrogen bonding recognition to develop a novel nanoparticle sensor for on-site and real-time detection of melamine in raw milk and infant formula without the aid of any advanced instrument. This colorimetric sensor could identify 20nM melamine in 1 min even with naked eye . Recently, increasing attention has been given to the development of aptamer GNP based colorimetric a say for the recognition of specific analytes (adenosine, cocaine, etc.) with high affinity and specificity [53-54].

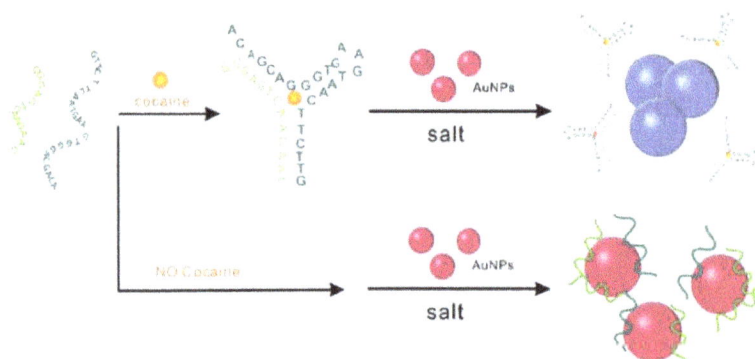

Visual cocaine detection using aptamer-GNPs as sensor

DNA Detection

Various approaches and techniques have been employed for the detection of DNA using GNPs as sensor [55-60]. A one-step homogeneous DNA detection method with high sensitivity was developed using GNPs coupled with dynamic light scattering (DLS) measurement by Dai et al. . This DLS-based assay is extremely easy to conduct and requires no additional separation and amplification steps. The detection limit is around 1 pM, four orders of magnitude better than that of light-absorption-based methods. Single base pair mismatched DNAs can be readily discriminated from perfectly matched target DNAs using this assay.

The oligonucleotide-mediated GNP aggregation process has been extensively used for the colorimetric screening of DNA binders and triplex DNA binders [60-61]. Furthermore, attractive sensors based on aptamer conjugated GNPs have been developed for the detection of a wide range of targets with high affinity and specificity [62-72]. An example is adenosine sensing, using an adenosine specific aptamer .The system contained two kinds of DNA-functionalized GNPs and a linker DNA. The linker DNA contained an adenosine aptamer fragment and an extension. In the presence of adenosine the nanoparticles are disaggregated, with a concomitant purple-to-red color change .

Protein Analysis

GNPs conjugated to antibodies have been widely used in the field of light and electron microscopy, for visualizing proteins in biological samples [73-74]. The use of GNPs for protein analysis/

detection is also a very interesting research field. In the past decade, GNP/protein conjugates have found increasing application as bioanalytical, diagnostic and/or immuno- histochemical probes [3, 5, 6, 75-83]. An optical resonance enhanced absorption based near- field immunochip bio-sensor for allergen detection was developed by Maier et. al.. Using this approach, a novel, simple, and rapid colorimetric solid-phase immunoassay on a planar chip substrate was realized in direct and sandwich assay formats, with a detection system that does not require any instrumentation for readout. In particular, semi quantitative immunochemical responses are directly visible to the naked eye of the analyst.

The combination of optical and electrochemical properties of GNPs with different detection techniques has been demonstrated in protein determination [84-89]. Ambrosi and co- workers synthesized a novel double codified nanolabel (DC-GNP) based on GNPs modified with an anti-human IgG peroxidase (HRP) conjugated antibody that allows enhanced spectrophotometric and electrochemical detection of antigen human IgG as a model protein .

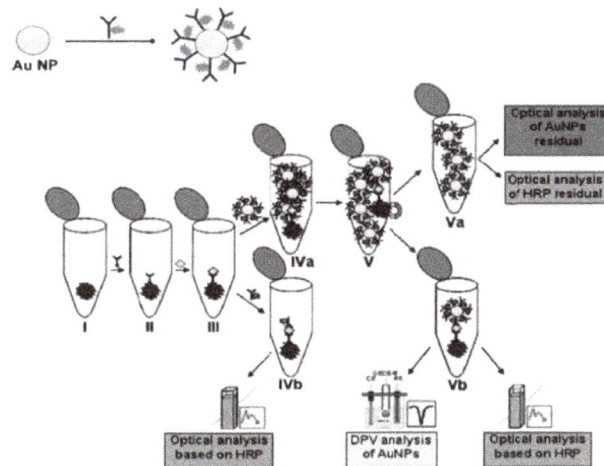

Another attractive sensor approach for protein detection uses aptamer conjugated GNPs that can bind target proteins with high affinity and specificity [84-87]. For example, Dong and co-workers have developed a kind of colorimetric sensor for detection of a-thrombin based on GNPs through the interaction of the aptamer to thrombin. The aptasensor has high sensitivity and selectivity and is able to detect targets in complex biological samples such as human plasma .

Enzymatic Activity Assay

Development of highly sensitive assay for determination of enzymatic activities and kinetic parameters is important in the fabrication of novel pharmaceuticals and medical diagnostic devices. Currently, the protein detection methods have largely been dominated by enzyme-linked immunological assay (known as ELISA) which is limited by the requirement of producing high-quality antibodies and is time consuming. For addressing these limitations, scientists have developed numerous novel detection strategies that incorporate the optical and electronic properties of GNPs [90-109]. Generally, these procedures circumvent the need for radioactive or fluorescent labeling usually required for assaying enzymatic activities. Some enzymes (protein kinases and proteases) responsive nanoparticle systems have been reviewed recently by Ghadiali and Stevents . The potential application of GNPs as colorimetric indicators to evaluate enzymatic activity and to screen enzyme inhibitors has been reported by Mirkin and co-workers . In principle, this method can be used to screen libraries of inhibitors of endonucleases in high through put fashion by using either the naked eye or a simple colorimetric reader.

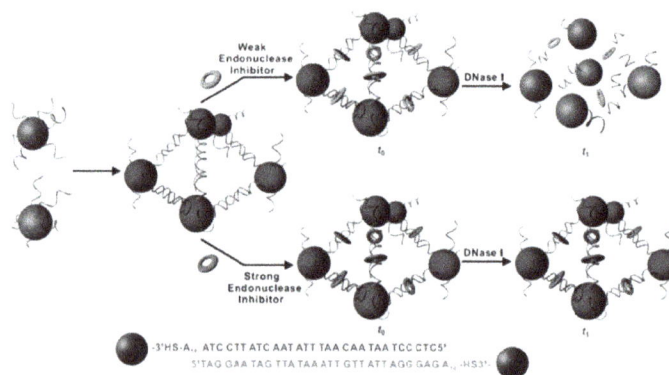

Illustration of the aggregation and dissociation of the DNA–GNP probes used in the colorimetric screening of endonuclease inhibitors.

ELISA

The enzyme-linked immunosorbent assay (ELISA) is a test that uses antibodies and color change to identify a substance.

ELISA is a popular format of "wet-lab" type analytic biochemistry assay that uses a solid-phase enzyme immunoassay (EIA) to detect the presence of a substance, usually an antigen, in a liquid sample or wet sample.

The ELISA has been used as a diagnostic tool in medicine and plant pathology, as well as a quality-control check in various industries.

Antigens from the sample are attached to a surface. Then, a further specific antibody is applied over the surface so it can bind to the antigen. This antibody is linked to an enzyme, and, in the final step, a substance containing the enzyme's substrate is added. The subsequent reaction produces a detectable signal, most commonly a color change in the substrate.

Performing an ELISA involves at least one antibody with specificity for a particular antigen. The sample with an unknown amount of antigen is immobilized on a solid support (usually a polystyrene microtiter plate) either non-specifically (via adsorption to the surface) or specifically (via capture by another antibody specific to the same antigen, in a "sandwich" ELISA). After the antigen is immobilized, the detection antibody is added, forming a complex with the antigen. The detection antibody can be covalently linked to an enzyme, or can itself be detected by a secondary antibody that is linked to an enzyme through bioconjugation. Between each step, the plate is typically washed with a mild detergent solution to remove any proteins or antibodies that are non-specifically bound. After the final wash step, the plate is developed by adding an enzymatic substrate to produce a visible signal, which indicates the quantity of antigen in the sample.

Of note, ELISA can perform other forms of ligand binding assays instead of strictly "immuno" assays, though the name carried the original "immuno" because of the common use and history of development of this method. The technique essentially requires any ligating reagent that can be immobilized on the solid phase along with a detection reagent that will bind specifically and use an enzyme to generate a signal that can be properly quantified. In between the washes, only the ligand and its specific binding counterparts remain specifically bound or "immuno-sorbed" by antigen-antibody interactions to the solid phase, while the nonspecific or unbound components are washed away. Unlike other spectrophotometric wet lab assay formats where the same reaction well (e.g. a cuvette) can be reused after washing, the ELISA plates have the reaction products immunosorbed on the solid phase which is part of the plate, and so are not easily reusable.

Principle

As an analytic biochemistry assay, ELISA involves detection of an "analyte" (i.e. the specific substance whose presence is being quantitatively or qualitatively analyzed) in a liquid sample by a method that continues to use liquid reagents during the "analysis" (i.e. controlled sequence of biochemical reactions that will generate a signal which can be easily quantified and interpreted as a measure of the amount of analyte in the sample) that stays liquid and remains inside a reaction chamber or well needed to keep the reactants contained; It is opposed to "dry lab" that can use dry strips – and even if the sample is liquid (e.g. a measured small drop), the final detection step in "dry" analysis involves reading of a dried strip by methods such as reflectometry and does not need a reaction containment chamber to prevent spillover or mixing between samples.

As a heterogenous assay, ELISA separates some component of the analytical reaction mixture by adsorbing certain components onto a solid phase which is physically immobilized. In ELISA, a liquid sample is added onto a stationary solid phase with special binding properties and is followed by multiple liquid reagents that are sequentially added, incubated and washed followed by some optical change (e.g. color development by the product of an enzymatic reaction) in the final liquid in the well from which the quantity of the analyte is measured. The qualitative "reading" usually based on detection of intensity of transmitted light by spectrophotometry, which involves quantitation of transmission of some specific wavelength of light through the liquid (as well as the transparent bottom of the well in the multiple-well plate format). The sensitivity of detection depends on amplification of the signal during the analytic reactions. Since enzyme reactions are very well known amplification processes, the signal is generated by enzymes which are linked to the detection reagents in fixed proportions to allow accurate quantification – thus the name "enzyme linked".

The analyte is also called the ligand because it will specifically bind or ligate to a detection reagent, thus ELISA falls under the bigger category of ligand binding assays. The ligand-specific binding reagent is "immobilized", i.e., usually coated and dried onto the transparent bottom and sometimes also side wall of a well (the stationary "solid phase'/"solid substrate" here as opposed to solid microparticle/beads that can be washed away), which is usually constructed as a multiple-well plate known as the "ELISA plate". Conventionally, like other forms of immunoassays, the specificity of antigen-antibody type reaction is used because it is easy to raise an antibody specifically against an antigen in bulk as a reagent. Alternatively, if the analyte itself is an antibody, its target antigen can be used as the binding reagent.

History

Before the development of the ELISA, the only option for conducting an immunoassay was radioimmunoassay, a technique using radioactively labeled antigens or antibodies. In radioimmunoassay, the radioactivity provides the signal, which indicates whether a specific antigen or antibody is present in the sample. Radioimmunoassay was first described in a scientific paper by Rosalyn Sussman Yalow and Solomon Berson published in 1960.

Because radioactivity poses a potential health threat, a safer alternative was sought. A suitable alternative to radioimmunoassay would substitute a nonradioactive signal in place of the radioactive signal. When enzymes (such as horseradish peroxidase) react with appropriate substrates (such as ABTS or TMB), a change in color occurs, which is used as a signal. However, the signal has to be associated with the presence of antibody or antigen, which is why the enzyme has to be linked to an appropriate antibody. This linking process was independently developed by Stratis Avrameas and G. B. Pierce. Since it is necessary to remove any unbound antibody or antigen by washing, the antibody or antigen has to be fixed to the surface of the container; i.e., the immunosorbent must be prepared. A technique to accomplish this was published by Wide and Jerker Porath in 1966.

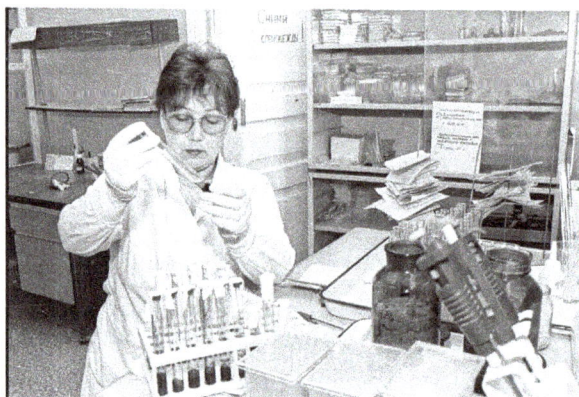

A paramedic assistant prepares analyses in an ELISA laboratory

In 1971, Peter Perlmann and Eva Engvall at Stockholm University in Sweden, and Anton Schuurs and Bauke van Weemen in the Netherlands independently published papers that synthesized this knowledge into methods to perform EIA/ELISA.

Traditional ELISA typically involves chromogenic reporters and substrates that produce some kind of observable color change to indicate the presence of antigen or analyte. Newer ELISA-like techniques use fluorogenic, electrochemiluminescent, and quantitative PCR reporters to create quantifiable signals. These new reporters can have various advantages, including higher sensitivities and multiplexing. In technical terms, newer assays of this type are not strictly ELISAs, as they are not "enzyme-linked", but are instead linked to some nonenzymatic reporter. However, given that the general principles in these assays are largely similar, they are often grouped in the same category as ELISAs.

In 2012 an ultrasensitive, enzyme-based ELISA test using nanoparticles as a chromogenic reporter was able to give a naked-eye colour signal from the detection of mere attograms of analyte. A blue color appears for positive results and red color for negative. Note that this detection only can confirm the presence or the absence of analyte not the actual concentration.

Types

Direct ELISA

Virus Sample on Surface

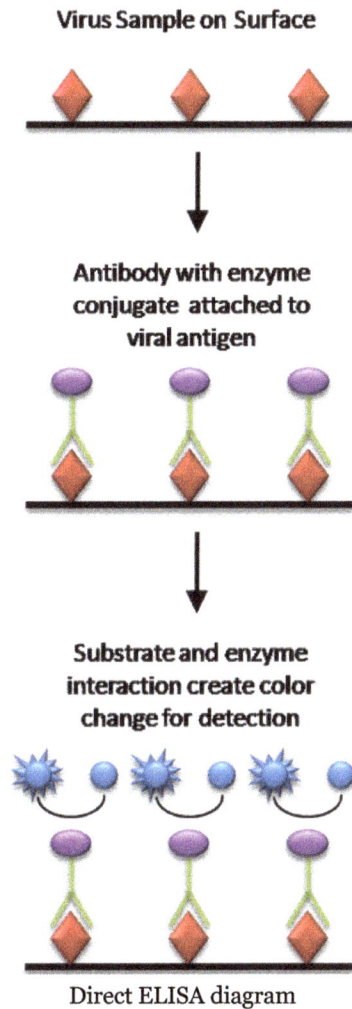

**Antibody with enzyme
conjugate attached to
viral antigen**

**Substrate and enzyme
interaction create color
change for detection**

Direct ELISA diagram

The steps of direct ELISA follows the mechanism below:

- A buffered solution of the antigen to be tested for is added to each well of a microtiter plate, where it is given time to adhere to the plastic through charge interactions.

- A solution of nonreacting protein, such as bovine serum albumin or casein, is added to well (usually 96-well plates) in order to cover any plastic surface in the well which remains uncoated by the antigen.

- The primary antibody with an attached (conjugated) enzyme is added, which binds specifically to the test antigen coating the well.

- A substrate for this enzyme is then added. Often, this substrate changes color upon reaction with the enzyme.

- The higher the concentration of the primary antibody present in the serum, the stronger the color change. Often, a spectrometer is used to give quantitative values for color strength.

The enzyme acts as an amplifier; even if only few enzyme-linked antibodies remain bound, the enzyme molecules will produce many signal molecules. Within common-sense limitations, the enzyme can go on producing color indefinitely, but the more antibody is bound, the faster the color will develop. A major disadvantage of the direct ELISA is the method of antigen immobilization is not specific; when serum is used as the source of test antigen, all proteins in the sample may stick to the microtiter plate well, so small concentrations of analyte in serum must compete with other serum proteins when binding to the well surface. The sandwich or indirect ELISA provides a solution to this problem, by using a "capture" antibody specific for the test antigen to pull it out of the serum's molecular mixture.

ELISA may be run in a qualitative or quantitative format. Qualitative results provide a simple positive or negative result (yes or no) for a sample. The cutoff between positive and negative is determined by the analyst and may be statistical. Two or three times the standard deviation (error inherent in a test) is often used to distinguish positive from negative samples. In quantitative ELISA, the optical density (OD) of the sample is compared to a standard curve, which is typically a serial dilution of a known-concentration solution of the target molecule. For example, if a test sample returns an OD of 1.0, the point on the standard curve that gave OD = 1.0 must be of the same analyte concentration as the sample.

The use and meaning of the names "direct ELISA" and "indirect ELISA" differs in the literature and on web sites depending on the context of the experiment. When the presence of an antigen is analyzed, the name "direct ELISA" refers to an ELISA in which only a labelled primary antibody is used, and the term "indirect ELISA" refers to an ELISA in which the antigen is bound by the primary antibody which then is detected by a labeled secondary antibody. In the latter case a sandwich ELISA is clearly distinct from an indirect ELISA. When the "primary" antibody is of interest, e.g. in the case of immunization analyses, this antibody is directly detected by the secondary antibody and the term "direct ELISA" applies to a setting with two antibodies.

Sandwich ELISA

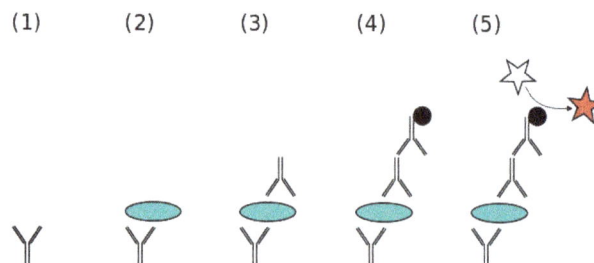

A sandwich ELISA. (1) Plate is coated with a capture antibody; (2) sample is added, and any antigen present binds to capture antibody; (3) detecting antibody is added, and binds to antigen; (4) enzyme-linked secondary antibody is added, and binds to detecting antibody; (5) substrate is added, and is converted by enzyme to detectable form.

A "sandwich" ELISA is used to detect sample antigen. The steps are:

1. A surface is prepared to which a known quantity of capture antibody is bound.

2. Any nonspecific binding sites on the surface are blocked.

3. The antigen-containing sample is applied to the plate, and captured by antibody.

4. The plate is washed to remove unbound antigen.

5. A specific antibody is added, and binds to antigen (hence the 'sandwich': the Ag is stuck between two antibodies). This primary antibody could also be in the serum of a donor to be tested for reactivity towards the antigen.

6. Enzyme-linked secondary antibodies are applied as detection antibodies that also bind specifically to the antibody's Fc region (nonspecific).

7. The plate is washed to remove the unbound antibody-enzyme conjugates.

8. A chemical is added to be converted by the enzyme into a color or fluorescent or electrochemical signal.

9. The absorbance or fluorescence or electrochemical signal (e.g., current) of the plate wells is measured to determine the presence and quantity of antigen.

The image to the right includes the use of a secondary antibody conjugated to an enzyme, though, in the technical sense, this is not necessary if the primary antibody is conjugated to an enzyme (which would be direct ELISA). However, the use of a secondary-antibody conjugate avoids the expensive process of creating enzyme-linked antibodies for every antigen one might want to detect. By using an enzyme-linked antibody that binds the Fc region of other antibodies, this same enzyme-linked antibody can be used in a variety of situations. Without the first layer of "capture" antibody, any proteins in the sample (including serum proteins) may competitively adsorb to the plate surface, lowering the quantity of antigen immobilized. Use of the purified specific antibody to attach the antigen to the plastic eliminates a need to purify the antigen from complicated mixtures before the measurement, simplifying the assay, and increasing the specificity and the sensitivity of the assay. A sandwich ELISA used for research often need validation because of the risk of false positive results.

Competitive ELISA

A third use of ELISA is through competitive binding. The steps for this ELISA are somewhat different from the first two examples:

1. Unlabeled antibody is incubated in the presence of its antigen (sample).

2. These bound antibody/antigen complexes are then added to an antigen-coated well.

3. The plate is washed, so unbound antibodies are removed. (The more antigen in the sample, the more Ag-Ab complexes are formed and so there are less unbound antibodies available to bind to the antigen in the well, hence "competition".)

4. The secondary antibody, specific to the primary antibody, is added. This second antibody is coupled to the enzyme.

5. A substrate is added, and remaining enzymes elicit a chromogenic or fluorescent signal.

6. The reaction is stopped to prevent eventual saturation of the signal.

Some competitive ELISA kits include enzyme-linked antigen rather than enzyme-linked antibody. The labeled antigen competes for primary antibody binding sites with the sample antigen (unlabeled). The less antigen in the sample, the more labeled antigen is retained in the well and the stronger the signal.

Commonly, the antigen is not first positioned in the well.

For the detection of HIV antibodies, the wells of microtiter plate are coated with the HIV antigen. Two specific antibodies are used, one conjugated with enzyme and the other present in serum (if serum is positive for the antibody). Cumulative competition occurs between the two antibodies for the same antigen, causing a stronger signal to be seen. Sera to be tested are added to these wells and incubated at 37 °C, and then washed. If antibodies are present, the antigen-antibody reaction occurs. No antigen is left for the enzyme-labelled specific HIV antibodies. These antibodies remain free upon addition and are washed off during washing. Substrate is added, but there is no enzyme to act on it, so a positive result shows no color change.

Applications

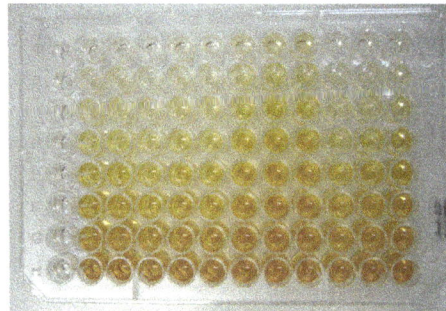

Human anti-IgG, double antibody sandwich ELISA

Because the ELISA can be performed to evaluate either the presence of antigen or the presence of antibody in a sample, it is a useful tool for determining serum antibody concentrations (such as with the HIV test or West Nile virus). It has also found applications in the food industry in detecting potential food allergens, such as milk, peanuts, walnuts, almonds, and eggs and as serological blood test for coeliac disease. ELISA can also be used in toxicology as a rapid presumptive screen for certain classes of drugs.

Enzyme-linked immunosorbent assay plate

The ELISA was the first screening test widely used for HIV because of its high sensitivity. In an ELISA, a person's serum is diluted 400 times and applied to a plate to which HIV antigens are attached. If antibodies to HIV are present in the serum, they may bind to these HIV antigens. The plate is then washed to remove all other components of the serum. A specially prepared "secondary antibody" — an antibody that binds to other antibodies — is then applied to the plate, followed by another wash. This secondary antibody is chemically linked in advance to an enzyme.

Thus, the plate will contain enzyme in proportion to the amount of secondary antibody bound to the plate. A substrate for the enzyme is applied, and catalysis by the enzyme leads to a change in color or fluorescence. ELISA results are reported as a number; the most controversial aspect of this test is determining the "cut-off" point between a positive and a negative result.

A cut-off point may be determined by comparing it with a known standard. If an ELISA test is used for drug screening at workplace, a cut-off concentration, 50 ng/ml, for example, is established, and a sample containing the standard concentration of analyte will be prepared. Unknowns that generate a stronger signal than the known sample are "positive." Those that generate weaker signal are "negative".

Dr Dennis E Bidwell and Alister Voller created the ELISA test to detect various kind of diseases, such as malaria, Chagas disease, and Johne's disease. ELISA tests also are used as in *in vitro* diagnostics in medical laboratories. The other uses of ELISA include:

- detection of *Mycobacterium* antibodies in tuberculosis

- detection of rotavirus in feces

- detection of hepatitis B markers in serum

- detection of enterotoxin of *E. coli* in feces

- detection of HIV antibodies in blood samples

Cellular Analysis

The ease of synthesis and functionalization, unique optical, properties, and biocompatibility of gold nanoparticles has recently sparked great interest in gold nanoparticles as a scaffold for cell-targeting studies [31,110–118]. Early and accurate detection of cancer often requires time consuming techniques and expensive instrumentation. To address these limitations, Tan's group developed a colorimetric assay for the direct detection of diseased cells . The assay uses aptamer-conjugated GNPs to combine the selectivity and af fi nity of aptamers and the spectroscopic advantages of GNPs to allow for the sensitive detection of cancer cells. Samples with the target cells exhibited a distinct color change while non-target samples did not elicit any change in color. The assay also showed excellent sensitivity with both the naked eye and based on absorbance measurements. In addition, the assay was able to differentiate between different types of target and control cells based on the aptamer used in the assay indicating the wide applicability of the assay for diseased cell detection. On the basis of these qualities aptamer- conjugated GNPs could become a powerful tool for point of care diagnostics.

GNPs are excellent platform for a diverse array of developing analytical methods and they have already been used for a wide range of applications both in chemical and biological research. The surface and core properties of these systems can be engineered for individual and multifold applications, including molecular recognition, chemical sensing and imaging. However, there are a number of critical issues that require addressing, including acute reproducible and reliable manufacturing methods/assays and long-term health effects of nanomaterials as well as scalability.

Route of Administration

A route of administration in pharmacology and toxicology is the path by which a drug, fluid, poison, or other substance is taken into the body. Routes of administration are generally classified by the location at which the substance is applied. Common examples include oral and intravenous administration. Routes can also be classified based on where the target of action is. Action may be topical (local), enteral (system-wide effect, but delivered through the gastrointestinal tract), or parenteral (systemic action, but delivered by routes other than the GI tract).

Classification

Routes of administration are usually classified by application location (or exposition). The route or course the active substance takes from application location to the location where it has its target effect is usually rather a matter of pharmacokinetics (concerning the processes of uptake, distribution, and elimination of drugs). Nevertheless, some routes, especially the transdermal or transmucosal routes, are commonly referred to *routes of administration*. The location of the target effect of active substances are usually rather a matter of pharmacodynamics (concerning e.g. the physiological effects of drugs). Furthermore, there is also a classification of routes of administration that basically distinguishes whether the effect is local (in "topical" administration) or systemic (in "enteral" or "parenteral" administration).

Application Location

Gastrointestinal/Enteral

Administration through the gastrointestinal tract is sometimes termed *enteral or enteric administration* (literally meaning 'through the intestines'). *Enteral/enteric administration* usually includes *oral* (through the mouth) and *rectal* (into the rectum) administration, in the sense that these are taken up by the intestines. However, uptake of drugs administered orally may also occur already in the stomach, and as such *gastrointestinal* (along the gastrointestinal tract) may be a more fitting term for this route of administration. Furthermore, some application locations often classified as *enteral*, such as sublingual (under the tongue) and sublabial or buccal (between the cheek and gums/gingiva), are taken up in the proximal part of the gastrointestinal tract without reaching the intestines. Strictly enteral administration (directly into the intestines) can be used for systemic administration, as well as local (sometimes termed topical), such as in a contrast enema, whereby contrast media is infused into the intestines for imaging. However, for the purposes of classification based on location of effects, the term enteral is reserved for substances with systemic effects.

Many drugs as tablets, capsules, or drops are taken orally. Administration methods directly into the stomach include those by gastric feeding tube or gastrostomy. Substances may also be placed into the small intestines, as with a duodenal feeding tube and enteral nutrition. Enteric coated tablets are designed to dissolve in the intestine, not the stomach, because the drug present in the tablet causes irritation in the stomach.

The rectal route is an effective route of administration for many medications, especially those used at the end of life. The walls of the rectum absorb many medications quickly and effectively. Medications delivered to the distal one-third of the rectum at least partially avoid the "first pass effect" through the liver, which allows for greater bio-availability of many medications than that of the oral route. Rectal mucosa is highly vascularized tissue that allows for rapid and effective absorption of medications. In hospice care, a specialized rectal catheter, designed to provide comfortable and discreet administration of ongoing medications provides a practical way to deliver and retain liquid formulations in the distal rectum, giving health practitioners a way to leverage the established benefits of rectal administration.

Central Nervous System

- epidural (synonym: peridural) (injection or infusion into the epidural space), e.g. epidural anesthesia

- intracerebral (into the cerebrum) direct injection into the brain. Used in experimental research of chemicals and as a treatment for malignancies of the brain. The intracerebral route can also interrupt the blood brain barrier from holding up against subsequent routes.

- intracerebroventricular (into the cerebral ventricles) administration into the ventricular system of the brain. One use is as a last line of opioid treatment for terminal cancer patients with intractable cancer pain.

Other Locations

- epicutaneous or topical (application onto the skin). It can be used both for local effect as in allergy testing and typical local anesthesia, as well as systemic effects when the active substance diffuses through skin in a transdermal route.

- Sublingual and buccal medication administration is a way of giving someone medicine orally (by mouth). Sublingual administration is when medication is placed under the tongue to be absorbed by the body. The word "sublingual" means "under the tongue." Buccal administration involves placement of the drug between the gums and the cheek. These medications can come in the form of tablets, films, or sprays.

- extra-amniotic administration, between the endometrium and fetal membranes

- nasal administration (through the nose) can be used for topically acting substances, as well as for insufflation of e.g. decongestant nasal sprays to be taken up along the respiratory tract. Such substances are also called *inhalational*, e.g. inhalational anesthetics.

- intraarterial (into an artery), e.g. vasodilator drugs in the treatment of vasospasm and thrombolytic drugs for treatment of embolism

- intraarticular, into a joint space. Used in treating osteoarthritis

- intracardiac (into the heart), e.g. adrenaline during cardiopulmonary resuscitation (no longer commonly performed)

- Intracavernous injection, an injection into the base of the penis

- intradermal, (into the skin itself) is used for skin testing some allergens, and also for mantoux test for tuberculosis

- Intralesional (into a skin lesion), is used for local skin lesions, e.g. acne medication

- intramuscular (into a muscle), e.g. many vaccines, antibiotics, and long-term psychoactive agents. Recreationally the colloquial term 'muscling' is used.

- intraocular, into the eye, e.g., some medications for glaucoma or eye neoplasms

- intraosseous infusion (into the bone marrow) is, in effect, an indirect intravenous access because the bone marrow drains directly into the venous system. This route is occasionally used for drugs and fluids in emergency medicine and pediatrics when intravenous access is difficult. Recreationally the colloquial term 'boning' is used.

- intraperitoneal, (infusion or injection into the peritoneum) e.g. peritoneal dialysis

- intrathecal (into the spinal canal) is most commonly used for spinal anesthesia and chemotherapy

- Intrauterine

- Intravaginal administration, in the vagina

- intravenous (into a vein), e.g. many drugs, total parenteral nutrition

- Intravesical infusion is into the urinary bladder.

- intravitreal, through the eye

- subcutaneous (under the skin), e.g. insulin. Skin popping is a slang term that includes this method of administration, and is usually used in association with recreational drugs.

- transdermal (diffusion through the intact skin for systemic rather than topical distribution), e.g. transdermal patches such as fentanyl in pain therapy, nicotine patches for treatment of addiction and nitroglycerine for treatment of angina pectoris.

- Transmucosal (diffusion through a mucous membrane), e.g. insufflation (snorting) of cocaine, sublingual, i.e. under the tongue, sublabial, i.e. between the lips and gingiva, nitroglycerine, vaginal suppositories

Local or Systemic Effect

Routes of administration can also basically be classified whether the effect is local (in topical administration) or systemic (in enteral or parenteral administration):

- *topical*: local effect, substance is applied directly where its action is desired. Sometimes, however, the term *topical* is defined as applied to a localized area of the body or to the surface of a body part, without necessarily involving target effect of the substance, making the classification rather a variant of the classification based on application location.

- *enteral*: desired effect is systemic (non-local), substance is given via the digestive tract.

- *parenteral*: desired effect is systemic, substance is given by routes other than the digestive tract.

Topical

- epicutaneous (application onto the skin), e.g. allergy testing, typical local anesthesia

- inhalational, e.g. asthma medications

- enema, e.g. contrast media for imaging of the bowel

- ophthalmic drugs / eye drops (onto the conjunctiva), e.g. antibiotics for conjunctivitis

- otic drugs / ear drops - such as antibiotics and corticosteroids for otitis externa

- through mucous membranes in the body

Enteral

In this classification system, enteral administration is administration that involves any part of the gastrointestinal tract (enteric system) and has systemic effects:

- by mouth (orally), many drugs as tablets, capsules, or drops

- by gastric feeding tube, duodenal feeding tube, or gastrostomy, e.g., many drugs and enteral nutrition

- rectally, various drugs in suppository

Parenteral

Any route that is not enteral (*par-* + *enteral*), including:

- intravenous (into a vein), e.g. many drugs, total parenteral nutrition

- intra-arterial (into an artery), e.g. vasodilator drugs in the treatment of vasospasm and thrombolytic drugs for treatment of embolism

- intraosseous infusion (into the bone marrow) is, in effect, an indirect intravenous access because the bone marrow drains directly into the venous system. This route is now occasionally used for drugs and fluids in emergency medicine and pediatrics when intravenous access is difficult.

- intra-muscular

- intracerebral (into the brain parenchyma)

- intracerebroventricular (into cerebral ventricular system)

- intrathecal (an injection into the spinal canal)

- subcutaneous (under the skin), e.g. a hypodermoclysis

Factors Governing Choice of Routes of Drug Administration

The reason for choice of routes of drug administration are governing by various factors. Such as :

- physical and chemical properties of the drug. Here there physical properties of drug are solid , liquid and gas. And chemical properties of drug are solubility, stability, pH, irritancy etc.

- Site of desired action. Here the action of drug may be localised and approachable or generalised and non approachable.

- Rate of extent of absorption of the drug from different routes.

- effect of digestive juices and first phase of metabolism.

- Condition of the patient.

Oral

The oral route is generally the most convenient and carries the lowest cost. However, some drugs can cause gastrointestinal tract irritation. For drugs that come in delayed release or time-release formulations, breaking the tablets or capsules can lead to more rapid delivery of the drug than intended.

Topical

By delivering drugs almost directly to the site of action, the risk of systemic side effects is reduced. However, skin irritation may result, and for some forms such as creams or lotions, the dosage is difficult to control.

Sublingual

This method refers to the pharmacological route of administration by which drugs diffuse into the blood through tissues under the tongue. Many drugs are designed for sublingual administration, including cardiovascular drugs, steroids, barbiturates, opioid analgesics with poor gastrointestinal bioavailability, enzymes and, increasingly, vitamins and minerals.

Inhalation

Inhaled medications can be absorbed quickly, and act both locally and systemically. Proper technique with inhaler devices is necessary to achieve the correct dose. Some medications can have an unpleasant taste or irritate the mouth.

Inhalation by smoking a substance is likely the most rapid way to deliver drugs to the brain, as the substance travels directly to the brain without being diluted in the systemic circulation. The severity of dependence on psychoactive drugs tends to increase with more rapid drug delivery.

Injection

The term injection encompasses intravenous (IV), intramuscular (IM), and subcutaneous (SC) administration.

Injections act rapidly, with onset of action in 15–30 seconds for IV, 10–20 minutes for IM, and 15–30 minutes for SC. They also have essentially 100% bioavailability, and can be used for drugs that are poorly absorbed or ineffective when given orally. Some medications, such as certain antipsychotics, can be administered as long-acting intramuscular injections. Ongoing IV infusions can be used to deliver continuous medication or fluids.

Disadvantages of injections include potential pain or discomfort for the patient, and the requirement of trained staff using aseptic techniques for administration. However, in some cases patients are taught to self-inject, such as SC injection of insulin in patients with insulin-dependent diabetes mellitus. As the drug is delivered to the site of action extremely rapidly with IV injection, there is a risk of overdose if the dose has been calculated incorrectly, and there is an increased risk of side effects if the drug is administered too rapidly.

Uses

- Some routes can be used for topical as well as systemic purposes, depending on the circumstance. For example, inhalation of asthma drugs is targeted at the airways (topical effect), whereas inhalation of volatile anesthetics is targeted at the brain (systemic effect).

- On the other hand, identical drugs can produce different results depending on the route of administration. For example, some drugs are not significantly absorbed into the bloodstream from the gastrointestinal tract and their action after enteral administration is therefore different from that after parenteral administration. This can be illustrated by the action of naloxone (Narcan), an antagonist of opiates such as morphine. Naloxone counteracts opiate action in the central nervous system when given intravenously and is therefore used in the treatment of opiate overdose. The same drug, when swallowed, acts exclusively on the bowels; it is here used to treat constipation under opiate pain therapy and does not affect the pain-reducing effect of the opiate.

- Enteral routes are generally the most convenient for the patient, as no punctures or sterile procedures are necessary. Enteral medications are therefore often preferred in the treatment of chronic disease. However, some drugs can not be used enterally because their absorption in the digestive tract is low or unpredictable. Transdermal administration is a comfortable alternative; there are, however, only a few drug preparations that are suitable for transdermal administration.

- In acute situations, in emergency medicine and intensive care medicine, drugs are most often given intravenously. This is the most reliable route, as in acutely ill patients the absorption of substances from the tissues and from the digestive tract can often be unpredictable due to altered blood flow or bowel motility.

References

- Quantum Materials Corporation and the Access2Flow Consortium (2011). "Quantum materials corp achieves milestone in High Volume Production of Quantum Dots". Retrieved 7 July 2011

- Vo-Dinh, T.; Cullum, B. (2000). "Biosensors and biochips: Advances in biological and medical diagnostics". Fresenius' Journal of Analytical Chemistry. 366 (6–7): 540–551. doi:10.1007/s002160051549

- MFTTech (24 March 2015). "LG Electronics Partners with Dow to Commercialize LGs New Ultra HD TV with Quantum Dot Technology". Retrieved 9 May 2015

- Malenka, Eric J. Nestler, Steven E. Hyman, Robert C. (2009). Molecular neuropharmacology : a foundation for clinical neuroscience (2nd ed.). New York: McGraw-Hill Medical. ISBN 978-0-07-148127-4

- Pasco, Neil; Glithero, Nick. Lactose at-line biosensor 1st viable industrial biosensor? "Archived copy" (PDF). Archived from the original (PDF) on 8 February 2013. Retrieved 9 February 2016

- LaFave, T. Jr. (2013). "Correspondences between the classical electrostatic Thomson Problem and atomic electronic structure". Journal of Electrostatics. 71 (6): 1029–1035. doi:10.1016/j.elstat.2013.10.001

- "Oklahoma Administrative Code and Register > 195:20-1-3.1. Pediatric conscious sedation utilizing enteral methods (oral, rectal, sublingual)". Retrieved 2009-01-18

4

Introduction to Biogenic and Stealth Nanoparticles

The nanoparticles that are produced through the synthesis of biological systems are called biogenic nanoparticles. As the process does not use any chemical methods, it falls under the category of green synthesis. Stealth particles are those nanoparticles that can avoid immune recognition and reach a specified target. This section has been carefully written to provide an easy understanding of the varied facets of nanobiotechnology.

Biogenic Nanoparticles

Nanoparticles synthesized using biological systems (eg. Microbes, fishes, plants etc.) are referred to as biogenic nanoparticles. The process of synthesis of the nanostructures by organisms is also known as biomineralization. As the process involves no harsh chemicals and solvent systems, this method of nanoparticle synthesis is classed as a 'Green synthesis'. These nanoparticles possess the advantage of having uniform size and shape. Moreover, they have also been found to possess better stability owing to stabilization by proteins and other biomolecules from the organism. The biogenic nanoparticles may be sequestered in separate intracellular compartments inside the cells and are referred to as intracellular biogenic particles. Separation of these nanoparticles requires disruption of the cells. On the other hand, some organisms synthesize these nanoparticles outside the cell or send them outside post-synthesis. Such nanoparticles are referred to as extracellular biogenic nanoparticles.

Why do organisms synthesize nanoparticles? One of the major reasons involved in such synthesis is detoxification. The major route to shield the cell(s) from the toxicity of certain soluble ionic species such as metal ions is to convert them into insoluble forms through reduction or precipitation.

In general, four major detoxification strategies have been identified in organisms.

(a) Modifications in the cellular transport mechanisms to restrict entry of the toxic ions into the cell

(b) Sequestration of the toxic species within the cell (intracellular sequestration) or outside the cell (extracellular sequestration)

(c) Activation of energy-dependent efflux pathways to eliminate the toxic species

(d) Enzyme catalyzed oxidation or reduction of the toxic species to a less toxic form

In certain cases, the synthesis of nanoparticles is initiated to meet the cell's requirements for a functional component such as iron oxide or silica. Such synthesis occurs via oxidation and condensation-association routes. Some of these nanostructures have exquisite and complex morphology that is too difficult to be recreated using chemical routes of synthesis. Moreover, large-scale production of these biogenic nanoparticles can be easily accomplished and hence this area has been among the most widely explored areas in nanoscience and technology.

Synthesis of Biogenic Metal Nanoparticles

The synthesis of metal nanoparticles by biological systems is mainly a strategy employed to protect the system against the toxic effects of the soluble metal ions. Several species of microbes and plants exhibit tolerance to metal ions through this strategy. Both bacterial and fungal species have been employed for synthesizing metal nanoparticles. The most widely synthesized metal nanoparticles are silver and gold due to their widespread applications in many fields of science and technology. The metal nanoparticles synthesized using fungal species is referred to as 'mycogenic nanoparticles' while those synthesized using bacterial species is known as 'bacterioform nanoparticles'. Even higher plants have been shown to be effective in synthesizing silver and gold nanoparticles and the amount of reports available in this topic is too large to be summarized.

Did you know?

The huge number of fungal species that have been used to synthesize different nanoparticles with excellent stability and size control has led to the genesis of a new field known as "Myconanotechnology". The cost-effective large-scale production of metallic, semiconductor and metal oxide nanoparticles is being developed for various applications in this rapidly expanding field.

Some of the major organisms that have been employed to synthesize nanoparticles are listed in the following Table.

Table: Biological systems used for synthesizing nanoparticles

Name of the biological species	Nanoparticle	Size (nm)
Acalypha indica	Silver	20-30
Actinobacter spp.	Magnetite	
Aspergillus clavatus	Silver	10-25
Aspergillus flavus	Silver	8.92±1.61
Aspergillus fumigatus	Silver	5-25
Aspergillus niger	Silver	20
Apiin extracted from henna leaves	Silver & Gold	39; 7.5-65 (respectively)
Avena sativa (oat)	Gold	5-20 (pH 3-4) & 25-85 (pH 2)
Azadirachta indica	Silver	50
Brassica juncea (mustard)	Silver	2-35
Bacillus licheniformis	Silver	50
Bacillus licheniformis (culture supernatant)	Silver	50

Bacillus megaterium	Silver	46.9
Bacillus sp.	Silver	5-15
Bacillus subtilis	Silver	5-60
Bacillus subtilis (culture supernatant)	Silver	5-60
Brevibacterium casei	Silver	50
Candida glabrata	CdS	Not available
Carica Papaya	Silver	60-80
Cinnamomum camphora leaf	Silver	55-80
Citrus limon (lemon)	Silver	<50
Cladosporium cladosporioides	Silver	10-100
Clostridium thermoaceticum	CdS	Not available
Cochlospermum gossypium	Silver	3
Coriandrum satiyum leaf extract	Silver	26
Coriandrum sativum	Gold	6.75-57.91
Corynebacterium sp.	Silver	10-15
Cymbopogon flexuosus (lemongrass)	Gold	200-500
Cycas sp. (cycas)	Silver	2-6
Datura metel	Silver	16-40
Desmodium triflorum	Silver	5-20
Eclipta sp.	Silver	2-6
Enhydra fluctuans	Silver	100-400
Escherichia coli	Silver	1-100
Escherichia coli	CdS	Not available
Eucalyptus camaldulensis (river red gum)	Gold	1.25-17.5
Eucalyptus citriodora (neelagiri)	Silver	~20
Eucalyptus hybrida (safeda)	Silver	50-150
Euphorbia hirta	Silver	40-50
Ficus bengalensis (marri)	Silver	~20

Fusarium acuminatum	Silver	5-40
Fusarium oxysporum	Silver	5-50
Fusarium oxysporum and Verticillium sp.	Magnetite	Not available
Fusarium semitectum	Silver	10-60
Fusarium solani	Silver	5-35
Garcinia mangostana (mangosteen)	Silver	35
Geobacter sulfurreducens	Silver	200
Gliricidia sepium	Silver	10-50
Glycine max (soybean) leaf extract	Silver	25-100
Honey	Silver	4
Ipomoea aquatic	Silver	100-400
Jatropha curcas (seed extract)	Silver	15-50
Klebsiella pneumonia (culture supernatant)	Silver	50

Lactic acid bacteria	Silver	11.2
Lactobacillus Strains	Silver	500
Ludwigia adscendeous (ludwigia)	Silver	100-400
Magnetospirillium magneticum	Fe_3O_4, Fe_3S_4	Not available
Mentha piperita (peppermint)	Silver & Gold	5-30, 90; 150
Moringa oleifera	Silver	57
Morganella sp.	Silver	20±5
Murraya koenigii	Silver & Gold	10; 20
Nelumbo nucifera (lotus)	Silver	25-80
Ocimum sanctum (tulsi; root extract)	Silver	10
Ocimum sanctum (tulsi; leaf extract)	Silver	10-20
Penicillium brevicompactum WA 2315	Silver	23-105
Penicillium fellutanum	Silver	1-100
Phanerochaete chrysosporium	Silver	100
Phyllanthus amarus	Silver	18-38
Proteus mirabilis	Silver	10-20
Psidium guajava (guava)	Gold	25-30
Pseudomonas stutzeri AG259	Silver	200
Scutellaria barbata D. Don (Barbated skullcup)	Gold	5-30
Schizosaccharomyces pombe	CdS	Not available
Sesbania drummondii (leguminous)	Gold	6-20
Stapphylococcus aurens	Silver	1-100
Syzygium aromaticum (clove)	Gold	5-100
Syzygium cumini (jambul)	Silver	29-92
Terminalia catappa (almond)	Gold	10-35
Trichoderma asperellum	Silver	13-18
Trichoderma viride	Silver	5-40
Verticillium sp.	Silver	25±12

The differences in the reducing agents, stabilizing moieties as well as templating molecules between different species cause variations in the size, size distribution as well as crystal type of the nanoparticles.

Mechanism Involved in the Synthesis of Biogenic Metal Nanoparticles

Though different biogenic sources have been identified for synthesis of nanoparticles, the mechanism of synthesis is yet to be deciphered clearly. Many theories have been suggested to explain the mechanism of synthesis, especially for metal nanoparticles. As biogenic metal nanoparticles have been found to be stable for three months and more, it is evident that there should be a reducing agent as well as a stabilizing agent in the system that would facilitate formation and stabilization of the metal nanoparticles. The identification of these components from the complex cellular environment is extremely challenging. The nature of the reducing agent and stabilizing agent also varies with the species employed for the synthesis. In general, peptides, proteins, polyols and heterocyclic compounds have been suggested to have an important role in the synthesis of metal nanoparticles through an electron shuttle or charge capping mechanism.

Peptides have been investigated for their role as possible reducing agents as well as stabilizing agents in many microbial species. Interestingly, several reports have indicated that a single amino acid might not be as effective as a polypeptide sequence containing the same amino acid residue for the synthesis. The metal ion is first reduced to form metal nanoparticle, which acts as a nucleus for further growth of the metal crystal. The peptide is initially thought to adsorb to the surface of the metal nanoparticle clusters causing a localized reducing environment that results in the reduction of more metal ions at the interface between the peptide and the metal nuclei. As a result, nanoparticles with large size distributions are formed. Figure depicts a cartoon on the formation of a silver nanocrystal in the presence of peptide.

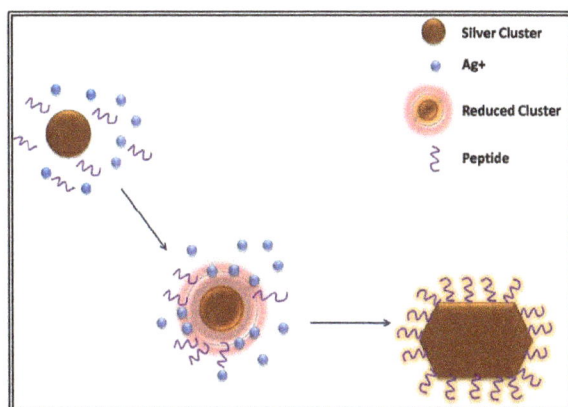

Formation of silver nanocrystals in the presence of peptide

Formation of twinned crystals could also occur in this process. A twinned crystal forms when two crystals have some similar lattice points in a symmetrical manner. A similar kind of mechanism has been proposed with proteins in plants and is referred to as the 'recognition-reduction-limited nucleation'. The positively charged metal ions interact electrostatically with negatively charged residues in the proteins followed by reduction resulting in nucleation at specific sites. The protein template further directs the growth of the crystals in a specific orientation.

Several amino acid residues have been implicated in the reduction and stabilization of silver and gold nanoparticles. These include arginine, cysteine, lysine, methionine, tyrosine and tryptophan. Tyrosine residue has been shown to reduce silver and gold ions under alkaline conditions. Figure gives the possible reaction involved in the formation of metal nanoparticles in the presence of tyrosine residue.

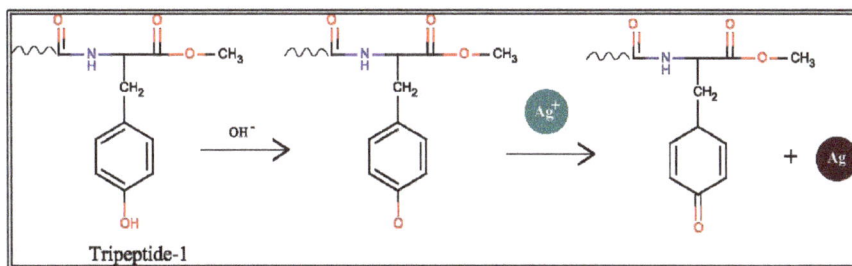

Role of tyrosine residue in the reduction of silver ions to form silver nanoparticles

The phenolic group of tyrosine gets converted into a semi-quinone during this process. The N-terminus of the peptide will interact with the metal nanoparticle conferring stability. Is there any evidence for this theory? Yes! It was found that an engineered viral protein template developed from the capsid of cowpea chlorotic mottle virus (CCMV) containing tyrosine residue was found to effectively synthesize gold nanoparticles.

Removal of the tyrosine residues resulted in no reduction reaction and hence no gold nanoparticle formation occurred.This observation strongly suggests that the presence of tyrosine residue is a key factor in the formation of metal nanoparticles. Figure shows a schematic representation of a viral protein containing tyrosine involved in the synthesis of metal nanoparticles.

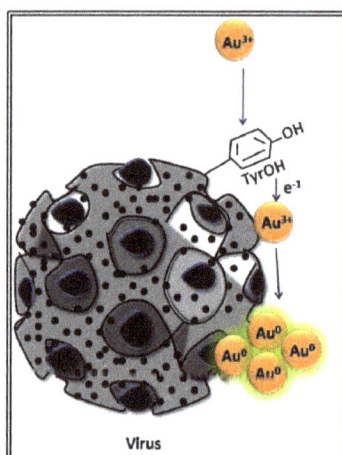

Viral capsid involved in the synthesis of gold nanoparticles

Tryptophan residue in peptides also has been shown to play a key role in the formation of metal nanoparticles. The tryptophan residue forms a transient tryptophyl radical after donating an electron to the metal ion during the reduction reaction. The highly unstable radical forms ditryptophan, a fluorescent dimer and kynurenine, another fluorescent molecule. This scheme of reactions is represented in Figure.

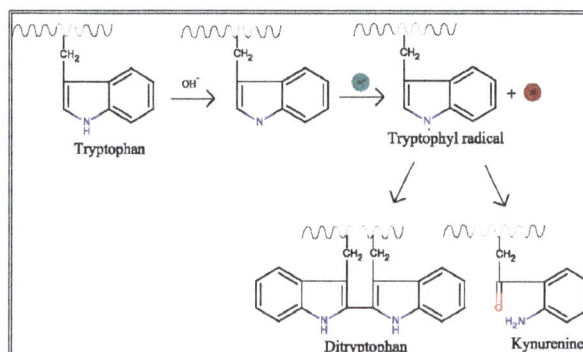

Role of tryptophan residue in the formation of metal nanoparticles

Biogenic Synthesis of Metal Nanoparticles

It has also been recognized now that in several bacterial species, the presence of the co- enzyme NADH along with the NADH-dependent nitrate reductase enzyme enables the reduction of the metal ions to the respective metal particle. The stability of the resultant particle is ensured by the presence of proteins as capping agents on the surface of the nanoparticle. The NADH-NAD$^+$ redox system acts as an electron shuttle between the enzyme and metal ion. Hence this mechanism is referred to as 'electron shuttle enzymatic metal reduction process'. Figure represents the sequence of events that occur in a NADH-dependent formation of silver nanoparticles.

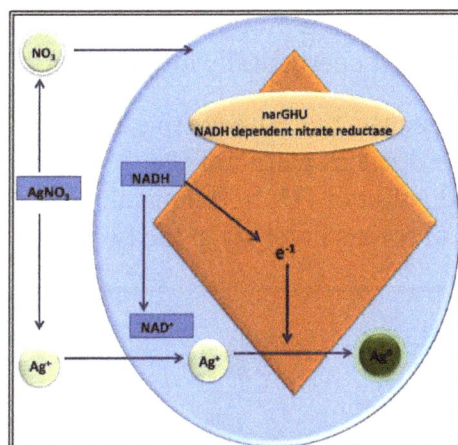

NADH-dependent electron shuttle enzymatic metal reduction process

The reaction did not occur in the absence of the NADH coenzyme. However, if the NADH was replaced by anthraquinone, the reduction reaction took place readily. Why? Anthraquinone is a redox mediator. In other words it can shuttle electrons from one molecular species to another. This implies that the presence of a redox mediator that could shuttle electrons between the enzyme and metal ion is essential in the reduction process leading to the formation of metal nanoparticles. Figure shows the scheme of events in the reduction process involving an anthraquinone moiety as the redox mediator.

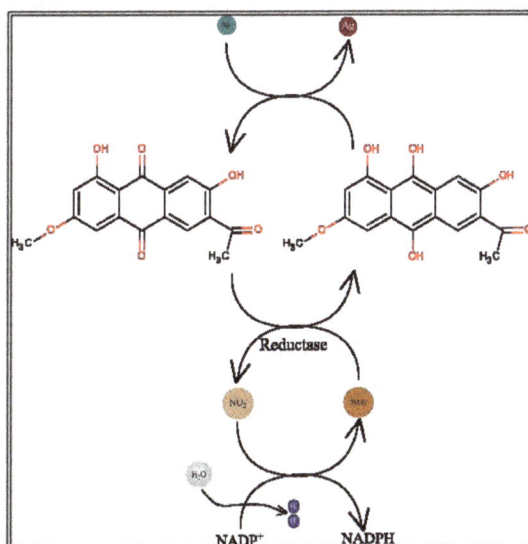

Anthraquinone mediated reduction of silver ions to form silver nanoparticles

In several species of bacteria utilizing NADPH/NADP+ redox system, charge capping of the resultant metal nanoparticles by the negatively charged phosphate from the coenzyme has been observed indicating an additional stabilizing role for the coenzyme.

Different enzymes have been identified in different species to have been involved in the formation of metal nanoparticles. In Lactobacillussp. mediated synthesis of metal nanoparticles, a NADH dependent lactate dehydrogenase was involved in the reduction reaction along with glutathione and thioredoxin systems.

In a fungus-mediated synthesis of platinum nanoparticles from chloroplatinic acid (H_2PtCl_4), a two stage, two-step reduction mechanism is proposed for the conversion of Pt^{4+} to Pt^0. This is because a single four-electron transfer is energetically unfavourable.

The enzyme cytoplasmic hydrogenase that is sensitive to oxygen catalyzes the conversion of tetravalent platinum (Pt^{4+}) to the divalent platinum (Pt^{2+}). An oxygen tolerant periplasmichydrogenase enzyme catalyzes the reduction of Pt^{2+} to the metal. This final step is inhibited by copper ions. Figure depicts the scheme of events involved in the two-step process.

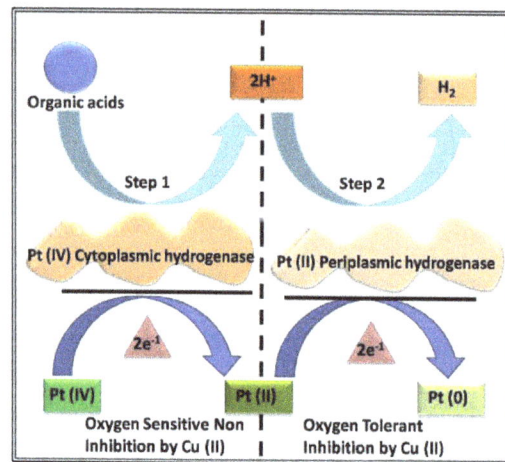

Two-step reduction reaction involved in the formation of platinum nanoparticles

The pH of the medium is an important factor that determines the efficiency of the reduction process. The type of enzyme dictates the optimum pH at which the catalysis is most facilitated. Higher pH facilitates creation of anionic centres that could act as nucleation sites for the positively charged metal ions by promoting electrostatic interactions.

Reducing sugars such as glucose found in the biological systems also have the ability to reduce metal ions to form metal nanoparticles. However, this reaction is pH dependent. The monosaccharide sugars predominantly exist in the cyclic structure. At alkaline pH, the ring opening occurs leading to the linear form that possesses the free aldehyde functional group. This aldehyde catalyses the reduction of the metal ions while the sugar molecule itself gets converted to the corresponding carboxylic acid. Figure depicts a reduction reaction involving glucose moiety.

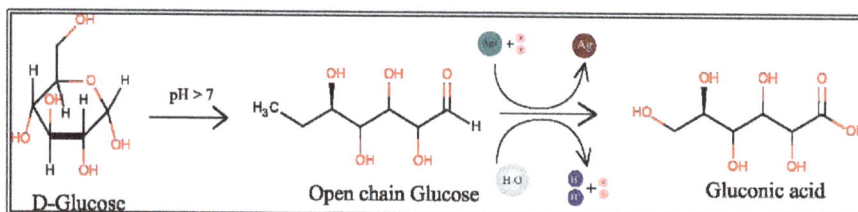

Formation of metal nanoparticles in the presence of glucose

This reaction also forms the basis of biochemical assays in laboratories to detect the presence of reducing sugars (Eg. Benedict's test, Tollen's silver mirror test etc.), though the particle sizes usually are not in the nanoscale owing to the absence of any stabilizers. Whether this reaction is dominant in biological systems to synthesize metal nanoparticles is still not very clear. It is also probable that in the cell, the proteins may serve as template to restrict the size of the particles formed to the nano-dimensions.

In higher plants, apart from enzymes, other molecules such as polyols, heterocyclic compounds and flavonoids may have a role in the synthesis of nanoparticles. Terpenoids such as citronellol and geraniol (found in rose oils) can reduce silver ions to silver while getting transformed to the corresponding carboxylic acids. Figure depict the transformation of silver ions to silver nanoparticles in the presence of geraniol.

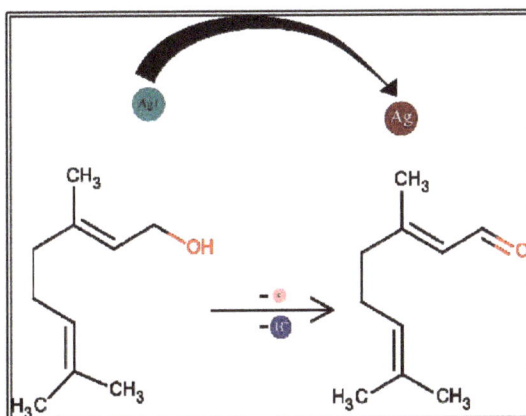

Formation of silver nanoparticles in the presence of a terpenoid, geraniol

Similarly, phyllanthin, a lignan component from the plant Phyllanthusamarus that possesses anti-microbial and anti-oxidant properties also has been found to convert gold and silver ions to their corresponding metal nanoparticles. The shape of the resultant nanoparticles was influenced by the concentration of phyllanthin. This can happen when the phyllanthin molecules interact with the surface of the nanoparticle clusters and form a cage-like template that dictates the shape of the crystal obtained.

In certain plants, catechol (1,2-dihydroxy benzene) under alkaline conditions converts metal ions to metal nanoparticles while it gets oxidized to protocatechaldehyde and finally protocatechuric acid. This reaction is shown in Figure.

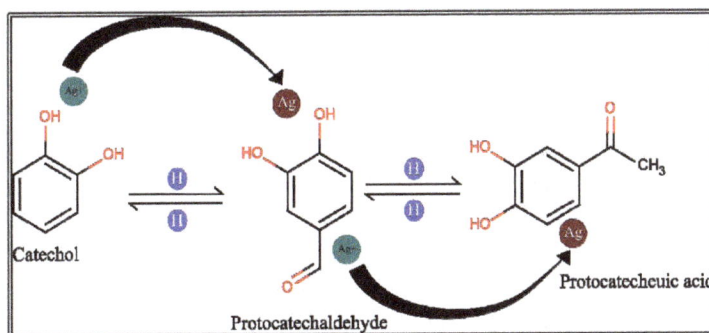

Formation of silver nanoparticles in the presence of catechol

Similarly, the flavonoid quercetin has been found to be involved in a two-step reduction leading to formation of metal nanoparticles. Ascorbic acid or vitamin C has also been implicated in the reduction of metal ions to metal nanoparticles. The ascorbic acid itself gets transformed into dehydroascorbic acid. However, the factors regulating this transformation are yet to be identified conclusively. Figure shows the reaction involved in the formation of silver nanoparticles by ascorbic acid.

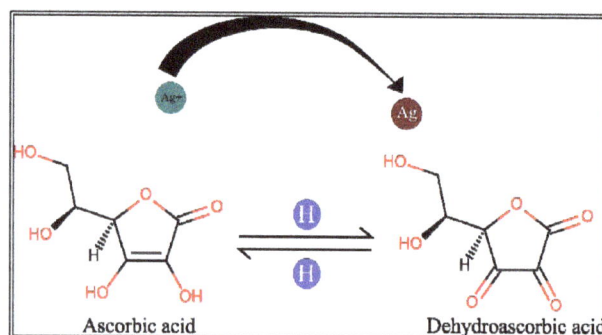

Formation of silver nanoparticles by ascorbic acid

Anthraquinones such as emodin, cyperquinone, remerin etc. undergo keto-enoltautomerism where the enol form acts as the reducing agent.

The Jatropacurcas plant, more famous for its biodiesel properties also has been found to effectively reduce silver and gold ions to the respective metal nanoparticles. This conversion is facilitated by the presence of two cyclic peptides – curcacyclin A (cyclic octapeptide) and curcacyclin B (cyclic nonapeptide). Figure shows the reaction involving the curcacyclin peptides leading to the formation of silver nanoparticles.

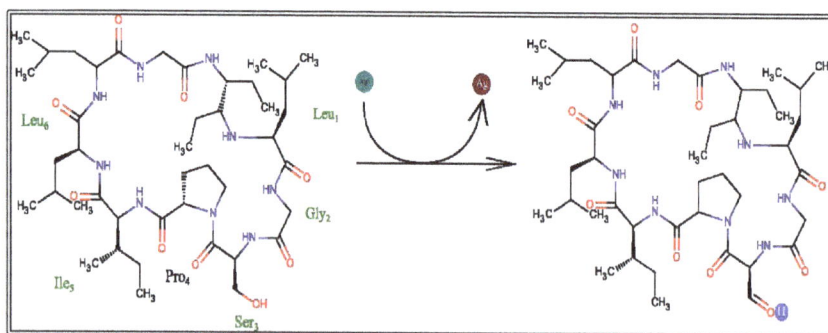

Formation of silver nanoparticles by curcacyclin peptides; Leu: leucine, Gly: glycine, Ser: serine, Ile: isoleucine

Though the exact mechanism of synthesis is not confirmed, it is proposed that the metal ions are accommodated in the core of the cyclic peptides. The amide groups in the peptide bond stabilize the ions. The reduction reaction is catalyzed by the alcohol group in the threonine residue in curcacyclin A and the serine residue in curcacyclin B.

Thus, it is evident that the biological system abounds with many types of reducing agents that can effectively convert the metal ions to metal nanoparticles. The binding of the metal ions is a key step in the synthesis. The stabilization and size control is effected by templates which are generally peptides or proteins.The amount of the peptide in the system could serve to stabilize the nanoparticles by adsorbing on the surface. However, if a large amount of peptide is present, then they tend to promote aggregation of the metal nanoparticles thereby contributing to increase in size and formation of unique shapes. Figure highlights this aspect using a peptide stabilized gold nanoparticle system as example.

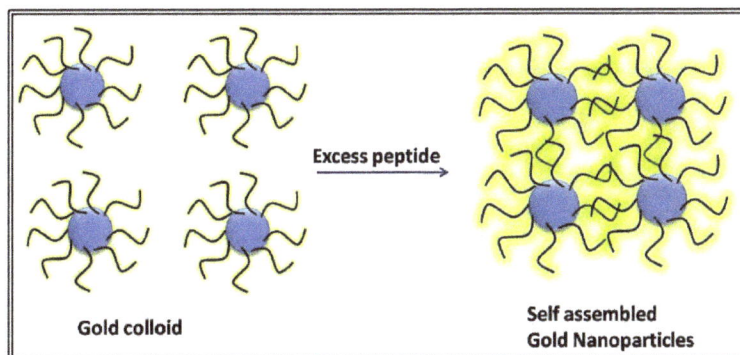

Influence of peptide concentration on size of the gold nano-structures obtained

Biogenic Synthesis of Quantum Dots

The formation of cadmium sulphide and zinc sulphide nanoparticles in microbes and plants has been mainly attributed to their detoxification attempts as both cadmium and zinc ions are toxic to these organisms in a concentration-dependent manner. Two categories of peptides have been reported to be involved in such detoxification processes. These are phytochelatins and metallothioneins. Though the name 'phytochelatin' suggests a plant origin, this peptide has also been found in animals. The phytochelatins are cysteine-rich oligopeptides that have excellent metal binding properties, especiallywith cadmium. Copper, nickel and zinc also bind with phytochelatins but at higher concentrations. The expression of these peptides is triggered by the presence of the heavy metals. Phytochelatins have repeated (γ-Glu-cys) sequences followed by an amino acid, mostly glycine. Other types containing alanine, serine, glutamate and glutamine as the final residue have also been identified in different species. The (γ-Glu-cys) sequence has an unusual peptide bond formed between the gamma carboxyl group of glutamate and alpha amino group of cysteine as in the case of glutathione. The number of (γ-Glu-cys) repeats in different phytochelatins varies from 2-11.

How does the presence of heavy metal ions trigger the expression of phytochelatins? The presence of metal ions in the biological system activates an enzyme called phytochelatin synthase, which utilizes glutathione from the cells to assemble phytochelatin. The next question that needs to be addressed is how does the phytochelatin aid in the formation of cadmium sulphide nanoparticles? The cadmium ions tend to bind to the thiol groups leading to the formation of a low molecular weight phytochelatin-cadmium complex. This peptide-metal ion complex is most likely transported by ABC (ATP-binding cassette) membrane transport proteins into a vacuole. This might be more common in plant cells. However, as vacuoles are not common in animal cells, the site of sequestration of these complexes in animal cells is yet to be identified.

Once inside the vacuoles, addition of more glutathione units to the complex is continued leading to increase in molecular weight. Simultaneously, sulphide ions are pumped into the vacuole. The source of these sulphide ions still is to be identified. But, it is widely believed thatsulphide genera-tion could arise due to the triggering of specific biochemical pathways by cadmium ions. In certain sulphate reducing bacteria, cysteine sulphinate is utilized to generate sulphide ions. These ions combine with the cadmium ions to form cadmium sulphide. The peptide template directs the size and shape of the resulting cadmium sulphide. Figure shows the schematic representation of the formation of cadmium sulphide using phytochelatin.

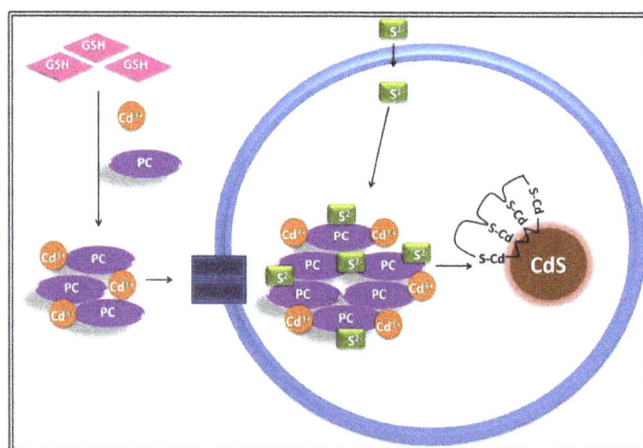

Formation of cadmium sulphide (CdS) by phytochelatin (PC); GSH represents reduced glutathione

Interestingly, the chelation of the cadmium ions to the phytochelatin units is stable only in weakly alkaline and alkaline pH (pH > 7). The pH in the vacuoles is about 5.5 where the phytochelatin–cadmium chelates loose their stability. Thus the cadmium ions now in the form of cadmium sul-phide are dissociated from the phytochelatin and remain in the vacuole while the free phytochela-tin units are transported away from the vacuole using specialized transport systems. Some species still retain a peptide coat, presumably an oligomer or monomeric unit from the phytochelatin unit,as a surface coating on the cadmium sulphide that stabilizes the CdS nanoparticles. The size of the nanoparticles depends on the length of the peptide main chain.

Extracellular synthesis of CdS nanoparticles has also been reported in certain fungal species. The cadmium ions present in the extracellular medium will be converted to cadmium sulphide crys-tals. The sulphide ions are obtained from the reduction of sulphate ions by the sulphatereductase enzymes secreted by the fungus. A co-enzyme siroheme is considered essential for this process.In some cases, the enzyme cysteine desulphydrase is implicated in the generation of sulphide ions from the amino acid cysteine.

Another alternate protein template that has been implicated in the synthesis of CdS nanoparticles is albumin. The globular albumin has a pocket 5- 10 nm in diameter that can accommodate cad-mium ions. The cysteine residues found in the albumin sequence can bind with the cadmium ions. This results in monodisperse and highly crystalline particles.

Metallothioneins are another class of cysteine-rich metal chelating proteins that may also have a role in the synthesis of sulphides of cadmium or copper or zinc, most likely in a manner similar to that of phytochelatins. Similarly, glutathione and cysteine are also said to play an important role in the synthesis of zinc sulphide and silver sulphide. However, the co-factors and other biomolecules involved in these processes are yet to be discovered.

Phytochelatin

Phytochelatins are oligomers of glutathione, produced by the enzyme phytochelatin synthase. They are found in plants, fungi, nematodes and all groups of algae including cyanobacteria. Phytochelatins act as chelators, and are important for heavy metal detoxification. They are abbreviated PC2 through PC11.

A mutant Arabidopsis thaliana lacking phytochelatin synthase is very sensitive to cadmium, but it grows just as well as the wild-type plant at normal concentrations of zinc and copper, two essential metal ions, indicating that phytochelatin is only involved in resistance to metal poisoning.

Because phytochelatin synthase uses glutathione with a blocked thiol group in the synthesis of phytochelatin, the presence of heavy metal ions that bind to glutathione causes the enzyme to work faster. Therefore, the amount of phytochelatin increases when the cell needs more phytochelatin to survive in an environment with high concentrations of metal ions.

Phytochelatin seems to be transported into the vacuole of plants, so that the metal ions it carries are stored safely away from the proteins of the cytosol.

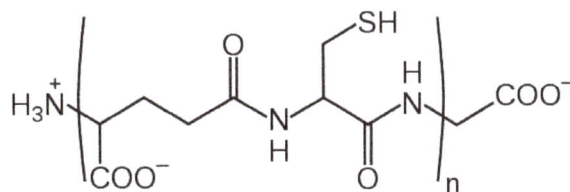

Chemical structure of phytochelatin. n = 2-11.

Related Peptides

There are groups of other peptides with a similar structure to phytochelatin, but where the last amino acid is not glycine:

Phytochelatin-like peptides

Phytochelatin-like peptides			
Type	Structure	Has been found in	Precursor
Phytochelatin	(γGlu-Cys)$_n$-Gly	many organisms	Glutathione
Homophytochelatin	(γGlu-Cys)$_n$-Ala	legumes	Homoglutathione
Desglycine phytochelatin	(γGlu-Cys)$_n$	maize, yeasts	
Hydroxymethyl-phytochelatin	(γGlu-Cys)$_n$-Ser	grasses	Hydroxymethylglutathione
iso-Phytochelatin (Glu)	(γGlu-Cys)$_n$-Glu	maize	Glutamylcysteinylglutamate
iso-Phytochelatin (Gln)	(γGlu-Cys)$_n$-Gln	horseradish	

History

Phytochelatin was first discovered in 1981 in fission yeast, and was named cadystin. It was then found in higher plants in 1985 and was named phytochelatin. In 1989 the biosynthetic enzyme, phytochelatin synthase, was discovered.

Metallothionein

Metallothionein (MT) is a family of cysteine-rich, low molecular weight (MW ranging from 500 to 14000 Da) proteins. They are localized to the membrane of the Golgi apparatus. MTs have the capacity to bind both physiological (such as zinc, copper, selenium) and xenobiotic (such as cadmium, mercury, silver, arsenic) heavy metals through the thiol group of its cysteine residues, which represent nearly 30% of its constituent amino acid residues.

MT was discovered in 1957 by Vallee and Margoshe from purification of a Cd-binding protein from horse (equine) renal cortex. MTs function is not clear, but experimental data suggest MTs may provide protection against metal toxicity, be involved in zinc and copper regulation, and provide protection against oxidative stress. There are four main isoforms expressed in humans: MT1 (subtypes A, B, E, F, G, H, L, M, X), MT2, MT3, and MT4. In the human body, large quantities are synthesised primarily in the liver and kidneys. Their production is dependent on availability of the dietary minerals such as zinc, copper, and selenium, as well as the amino acids histidine and cysteine.

Structure and Classification

MTs are present in a vast range of taxonomic groups, ranging from prokaryotes (such as the cyanobacteria *Synechococcus sp.*), protozoa (such as the ciliate *Tetrahymena* genera), plants (such as *Pisum sativum, Triticum durum, Zea mays*, or *Quercus suber*), yeast (such as *Saccharomyces cerevisiae* or *Candida albicans*), invertebrates (such as the nematode *Caenorhabditis elegans*, the insect *Drosophila melanogaster*, the mollusc *Mytilus edulis*, or the echinoderm *Strongylocentrotus purpuratus*) and vertebrates (such as the chicken *Gallus gallus*, or the mammalian *Homo sapiens* or *Mus musculus*).

The MTs from this diverse taxonomic range represent a high-heterogeneity sequence (regarding molecular weight and number and distribution of Cys residues) and do not show general homology; in spite of this, homology is found inside some taxonomic groups (such as vertebrate MTs).

From their primary structure, MTs have been classified by different methods. The first one dates from 1987, when Fowler *et al.*, established three classes of MTs: Class I, including the MTs which show homology with horse MT, Class II, including the rest of the MTs with no homology with horse MT, and Class III, which includes phytochelatins, Cys-rich enzymatically synthesised peptides. The second classification was performed by Binz and Kagi in 2001, and takes into account taxonomic parameters and the patterns of distribution of Cys residues along the MT sequence. It results in a classification of 15 families for proteinaceous MTs. Family 15 contains the plant MTs, which in 2002 have been further classified by Cobbet and Goldsbrough into 4 Types (1, 2, 3 and 4) depending on the distribution of their Cys residues and a Cys-devoid regions (called spacers) characteristic of plant MTs.

A table including the principal aspects of the two latter classifications is included.

Family	Name	Sequence pattern	Example
1	Vertebrate	K-x(1,2)-C-C-x-C-C-P-x(2)-C	*Mus musculus* MT1 MDPNCSCTTGGSCACAGSCKCKECKCTSCKKCCSCCPVG-CAKCAQGCVCKGSSEKCRCCA
2	Molluscan	C-x-C-x(3)-C-T-G-x(3)-C-x-C-x(3)-C-x-C-K	*Mytilus edulis* 10MTIV MPAPCNCIETNVCICDTGCSGEGCRCGDACKCSGADCKCS-GCKVVCKCSGSCACEGGCTGPSTCKCAPGCSCK

3	Crustacean	P-[GD]-P-C-C-x(3,4)-C-x-C	*Homarus americanus* MTH MPGPCCKDKCECAEGGCKTGCKCTSCRCAPCEKCTSGCK-CPSKDECAKTCSKPCKCCP
4	Echinoderms	P-D-x-K-C-V-C-C-x(5)-C-x-C-x(4)-C-C-x(4)-C-C-x(4,6)-C-C	*Strongylocentrotus purpuratus* SpMTA MPDVKCVCCKEGKECACFGQDCCKTGECCKDGTCCGICT-NAACKCANGCKCGSGCSCTEGNCAC
5	Diptera	C-G-x(2)-C-x-C-x(2)-Q-x(5)-C-x-C-x(2)D-C-x-C	*Drosophila melanogaster* MTNB MVCKGCGTNCQCSAQKCGDNCACNKDCQCVCKNGPKD-QCCSNK
6	Nematoda	K-C-C-x(3)-C-C	*Caenorhabditis elegans* MT1 MACKCDCKNKQCKCGDKCECSGDKCCEKYCCEEASEKKC-CPAGCKGDCKCANCHCAEQKQCGDKTHQHQGTAAAH
7	Ciliate	x-C-C-C-x ?	*Tetrahymena thermophila* MTT1 MDKVNSCCCGVNAKPCCTDPNSGCCCVSKTD-NCCKSDTKECCTGTGEGCKCVNCKCCKPQA-NCCCGVNAKPCCFDPNSGCCCVSKTNNCCKSD TKECCTGTGEGCKCTSCQCCKPVQQGCCCGDKAKACCTD-PNSGCCCSNKANKCCDATSKQECQTCQCCK
8	Fungal 1	C-G-C-S-x(4)-C-x-C-x(3,4)-C-x-C-S-x-C	*Neurospora crassa* MT MGDCGCSGASSCNCGSGCSCSNCGSK
9	Fungal 2	---	*Candida glabrata* MT2 MANDCKCPNGCSCPNCANGGCQCGDKCECKKQSCHGC-GEQCKCGSHGSSCHGSCGCGDKCECK
10	Fungal 3	---	*Candida glabrata* MT2 MPEQVNCQYDCHCSNCACENTCNCCAKPACACTNSAS-NECSCQTCKCQTCKC
11	Fungal 4	C-X-K-C-x-C-x(2)-C-K-C	*Yarrowia lipolytica* MT3 MEFTTAMLGASLISTTSTQSKHNLVNNCCCSSSTSESSM-PASCACTKCGCKTCKC
12	Fungal 5	---	*Saccharomyces cerevisiae* CUP1 MFSELINFQNEGHECQCQCGSCKNNEQCQKSCSCPT-GCNSDDKCPCGNKSEETKKSCCSGK
13	Fungal 6	---	*Saccharomyces cerevisiae* CRS5 TVKICDCEGECCKDSCHCGSTCLPSCSGGEKCKCDHSTG-SPQCKSCGEKCKCETTCTCEKSKCNCEKC
14	Procaryota	K-C-A-C-x(2)-C-L-C	*Synechococcus sp* SmtA MTTVTQMKCACPHCLCIVSLNDAIMVDGKPYCSEV-CANGTCKENSGCGHAGCGCGSA
15	Plant	[YFH]-x(5,25)-C-[SK-D]-C-[GA]-[SDPAT]-x(0,1)-C-x-[CYF]	
15.1	Plant MTs Type 1	C-X-C-X(3)- C-X-C-X(3)- C-X-C-X(3)-spacer-C-X-C-X(3)-C-X-C-X(3)- C-X-C-X(3)	*Pisum sativum* MT MSGCGCGSSCNCGDSCKCNKRSSGLSYSEMETTETVIL-GVGPAKIQFEGAEMSAASEDGGCKCGDNCTCDPCNCK
15.2	Plant MTs Type 2	C-C-X(3)-C-X-C-X(3)-C-X-C-X(3)- C-X-C-X(3)-spacer- C-X-C-X(3)- C-X-C-X(3)-C-X-C-X(3)	*Lycopersicon esculetum* MT MSCCGGNCGCGSSCKCGNGCGGCKMYPDM-SYTESSTTTETLVLGVGPEKTSFGAMEMGESPVAENGCK-CGSDCKCNPCTCSK

15.3	Plant MTs Type 3	---	*Arabidopsis thaliana* MT3 MSSNCGSCDCADKTQCVKKGTSYTFDIVETQESYKEAMIM-DVGAEENNANCKCKCGSSCSCVNCTCCPN
15.4	Plant MTs Type 4 or Ec	C-x(4)-C-X-C-X(3)-C-X(5)-C-X-C-X(9,11)-HTTCGCGEHC- X-C-X(20)-CSC-GAXCNCASC-X(3,5)	*Triticum aestivum* MT MGCNDKCGCAVPCPGGTGCRCTSARSDAAAGEHTTCG-CGEHCGCNPCACGREGTPSGRANRRANCSCGAACNCAS-CGSTTA
99	Phytochela-tins and other non-protein-aceous MT-like polypep-tides	---	*Schizosaccharomyces pombe* γEC-γEC-γECG

Secondary structure elements have been observed in several MTs SmtA from *Syneccochoccus*, mammalian MT3, Echinoderma SpMTA, fish *Notothenia coriiceps* MT, Crustacean MTH, but until this moment, the content of such structures is considered to be poor in MTs, and its functional influence is not considered.

Tertiary structure of MTs is also highly heterogeneous. While vertebrate, echinoderm and crustacean MTs show a bidominial structure with divalent metals as Zn(II) or Cd(II) (the protein is folded so as to bind metals in two functionally independent domains, with a metallic *cluster* each), yeast and procariotyc MTs show a monodominial structure (one domain with a single metallic *cluster*). Although no structural data is available for molluscan, nematoda and Drosophila MTs, it is commonly assumed that the former are bidominial and the latter monodominial. No conclusive data are available for Plant MTs, but two possible structures have been proposed: 1) a bidominial structure similar to that of vertebrate MTs; 2) a codominial structure, in which two Cys-rich domains interact to form a single metallic cluster.

Quaternary structure has not been broadly considered for MTs. Dimerization and oligomerization processes have been observed and attributed to several molecular mechanisms, including intermolecular disulfide formation, bridging through metals bound by either Cys or His residues on different MTs, or inorganic phosphate-mediated interactions. Dimeric and polymeric MTs have been shown to acquire novel properties upon metal detoxification, but the physiological significance of these processes has been demonstrated only in the case of prokaryotic Synechococcus SmtA. The MT dimer produced by this organism forms structures similar to zinc fingers and has Zn-regulatory activity.

Metallothioneins have diverse metal-binding preferences, which have been associated with functional specificity. As an example, the mammalian *Mus musculus* MT1 preferentially binds divalent metal ions (Zn(II), Cd(II),...), while yeast CUP1 is selective for monovalent metal ions (Cu(I), Ag(I),...). Strictly metal-selective MTs with metal-specific physiological functions were discovered by Dallinger et al. (1997) in pulmonate snails (Gastropoda, Mollusca). The Roman snail (*Helix pomatia*), for example, possesses a Cd-selective (CdMT) and a Cu-selective isoform (CuMT) involved in Cd detoxification and Cu regulation, respectively. While both isoforms contain unvaried numbers and positions of Cys residues responsible for metal ligation, metal selectivity is apparently achieved by sequence modulation of amino acid residues not directly involved in metal binding (Palacios et al. 2011).

A novel functional classification of MTs as Zn- or Cu-thioneins is currently being developed based on these functional preferences.

Yeast

Metallothioneins are characterized by an abundance of cysteine residues and a lack of generic secondary structure motifs. Yeast Metallothionein (MT) are also alternatively named, Copper metallothionein (CUP).

Function

This protein functions in primary metal storage, transport, and detoxification. More specifically, Yeast MT stores copper so therefore protects the cell against copper toxicity by tightly chelating copper ions.

Structure

For the first 40 residues in the protein the polypeptide wraps around the metal by forming two large parallel loops separated by a deep cleft containing the metal cluster.

Examples

Yeast MT can be found in the following:

- *Saccharomyces cerevisiae*

- *Neurospora crassa*

Function

Metal binding

Metallothionein has been documented to bind a wide range of metals including cadmium, zinc, mercury, copper, arsenic, silver, etc. Metallation of MT was previously reported to occur cooperatively but recent reports have provided strong evidence that metal-binding occurs via a sequential, noncooperative mechanism. The observation of partially metallated MT (that is, having some free metal binding capacity) suggest that these species are biologically important.

Metallothioneins likely participate in the uptake, transport, and regulation of zinc in biological systems. Mammalian MT binds three Zn(II) ions in its beta domain and four in the alpha domain. Cysteine is a sulfur-containing amino acid, hence the name "-thionein". However, the participation of inorganic sulfide and chloride ions has been proposed for some MT forms. In some MTs, mostly bacterial, histidine participates in zinc binding. By binding and releasing zinc, metallothioneins (MTs) may regulate zinc levels within the body. Zinc, in turn, is a key element for the activation and binding of certain transcription factors through its participation in the zinc finger region of the protein. Metallothionein also carries zinc ions (signals) from one part of the cell to another. When zinc enters a cell, it can be picked up by thionein (which thus becomes "metallothionein") and carried to another part of the cell where it is released to another organelle or protein. In this way the thionein-metallothionein becomes a key component of the zinc signaling system in cells. This system is particularly important in the brain, where zinc signaling is prominent both between and within nerve cells. It also seems to be important for the regulation of the tumor suppressor protein p53.

Control of Oxidative Stress

Cysteine residues from MTs can capture harmful oxidant radicals like the superoxide and hydroxyl radicals. In this reaction, cysteine is oxidized to cystine, and the metal ions which were bound to cysteine are liberated to the media. As explained in the *Expression and regulation* section, this Zn can activate the synthesis of more MTs. This mechanism has been proposed to be an important mechanism in the control of the oxidative stress by MTs. The role of MTs in oxidative stress has been confirmed by MT Knockout mutants, but some experiments propose also a prooxidant role for MTs.

Expression and Regulation

Metallothionein gene expression is induced by a high variety of stimuli, as metal exposure, oxidative stress, glucocorticoids, Vitamin D, hydric stress, etc. The level of the response to these inducers depends on the MT gene. MT genes present in their promotors specific sequences for the regulation of the expression, elements as metal response elements (MRE), glucocorticoid response elements (GRE), GC-rich boxes, basal level elements (BLE), and thyroid response elements (TRE).

Metallothionein and Disease

Cancer

Because MTs play an important role in transcription factor regulation, defects in MT function or expression may lead to malignant transformation of cells and ultimately cancer. Studies have found increased expression of MTs in some cancers of the breast, colon, kidney, liver, skin (melanoma), lung, nasopharynx, ovary, prostate, mouth, salivary gland, testes, thyroid and urinary bladder; they have also found lower levels of MT expression in hepatocellular carcinoma and liver adenocarcinoma.

There is evidence to suggest that higher levels of MT expression may also lead to resistance to chemotherapeutic drugs.

Autism

Heavy metal toxicity has been proposed as a hypothetical etiology of autism, and dysfunction of MT synthesis and activity may play a role in this. Many heavy metals, including mercury, lead, and arsenic have been linked to symptoms that resemble the neurological symptoms of autism. However, MT dysfunction has not specifically been linked to autistic spectrum disorders. A 2006 study, investigating children exposed to the vaccine preservative thiomersal, found that levels of MT and antibodies to MT in autistic children did not differ significantly from non-autistic children.

A low zinc to copper ratio has been seen as a biomarker for autism and suggested as an indication that the Metallothionein system has been affected.

Further, there is indication that the mother's zinc levels may affect the developing baby's immunological state that may lead to autism and could be again an indication that the Metallothionein system has been affected.

Magnetite Producing Organisms and Magnetosomes

It is now well known that biological systems can produce metal nanoparticles and even cadmium sulphide deposits. Is there any other form of inorganic nanoparticles that can be synthesized by biological organisms? We will explore about a category of magnetic nanoparticles that are biogenic in origin.

An interesting phenomenon observed on the motility of bacteria led to the discovery of magnetic deposits in several bacterial species. The main types of deposits in bacteria are magnetite and greigite. The magnetite (Fe_3O_4) deposits serve as internal compasses and as a result,the bacteria display specific directional movement with respect to the earth's magnetic field. This movement is known as magnetotaxis and such bacteria are called magnetotactic bacteria. The magnetotaxis is influenced by the presence of other ions in the environment. The most investigated magnetotactic bacteria are Magnetospirillium magnetotacticum and Magnetospirillium griphiswaldense.

How is this magnetite produced in bacteria? One of the first steps involved should be assimilation of iron in the form of Fe^{2+} or Fe^{3+} in the bacterial cell. The process is thought to be initiated by protein-assisted invagination of the cell membrane to form a vesicle. This may later mature to form a magnetosome. But this is yet to be confirmed. The transport of iron in the form of Fe^{2+} is aided by a hydrophobic transport protein MagA that is found in the lipid bilayer of the bacterial cell. It has both its N-terminus and C- terminus facing the intracellular direction. For transport it requires an ATPase pump nearby. The addition of ATP fuels the influx of protons into the cell through the ATPase. Once a threshold level of these protons is reached, they are pumped out by the MagA while simultaneously the Fe^{2+} ions are pumped into the system.Once inside the cell, the Fe^{2+} is oxidized to formlow densityhydrous ferric oxide followed by a dehydration step to form high density ferric oxide. Then, a small amount (about one third) of the ferric oxide is reduced and finally magnetite (Fe_3O_4) will be produced. These biomineralization reactions occur in unique lipid bilayer compartments known as 'magnetosomes'.

The lipid composition of magnetosomes shows presence of fatty acids, neutral lipids, sulpholipids, glycolipids, phosphatides, phosphatidyl ethanolamine and phosphatidyl serine. Several proteins that include MAM12, MAM22 and MAM28 have been found to be associated with the magnetosome. These contain charged amino acid residues that serve to establish electrostatic interactions with the cytoplasm and its components. The exact role of these proteins is yet to be identified. Most of the MAM polypeptides have been found to be tightly associated with the magnetosome and exhibit good resistance to detergents and proteases. Some of the MAM proteins show a homology with serine proteases and a few more with heat shock proteins. Several MAM proteins have similar sequences to cation diffusion facilitators implying a potential ion transport role. However, these are yet to confirmed by experimental findings.

It is believed that the magnetosome membrane not only enables transport of Fe^{2+}, it also may aid nucleation, crystallization and regulation of the pH and redox environment in the magnetosome.

Indeed, the formation of magnetite inside the magnetosome enables tight regulation of the nucleation and growth of the crystals. The size of the magnetite crystals can also be directed by the magnetosomes. Typical particle sizes range from 35-120 nm. In the bacterial cell, several crystals orient themselves to form linear chains. The number of chains as well as the number of crystals in each chain varies from species to species as well as their growth environment. Presence of many chains in a single bacterium will result in repulsive forces that push each chain to the corner of the cells resulting in a turgor pressure. Cubo-octahedral, hexagonal prismatic and bullet-shaped morphologies have been observed for magnetite crystals derived from different magnetotactic bacteria. The (111), (110) and (100) crystal faces have been identified in the magnetite crystals with the (111) crystal face being the most common. The alignment of the magnetite crystal chains usually is in the direction of the cytoskeletal filaments. Figure depicts the magnetic deposits inside a bacterial cell.

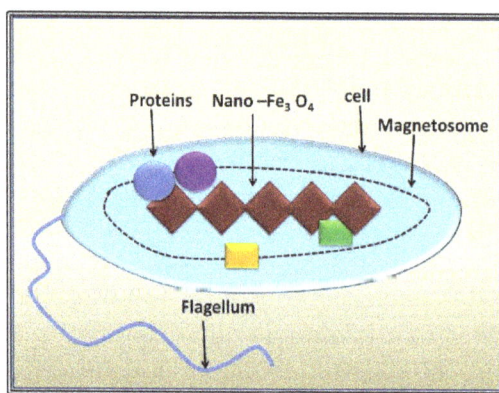

Cartoon of a bacteria containing a magnetosome

Greigite Synthesis

Several anaerobic bacteria in deep water bodies have been found to contain Fe_3S_4 (Greigite) deposits instead of Fe_3O_4 (magnetite). These bacterial are found well below the OATZ (Oxic-Anoxic Transition Zone). Bacteria found at the OATZ show both magnetite and Greigite deposits indicating that the presence of oxygen is essential for magnetite production. Below the OATZ, hydrogen sulphide content is very high and there is complete absence of molecular oxygen. Hence the microorganisms present in that zone have only Greigite deposits. Rectangular, prismatic as well as cubo-octahedral geometries have been observed in these biomineralized Greigite deposits. The size of each Greigite crystal ranges from 30-100 nm.

Biogenic Nanoparticles from Higher Organisms

Biomineralization of nanoparticles have been also discovered in multi-cellular organisms higher in the evolutionary tree. Deposits of magnetite have been found in the brain of salmon fish (Oncorhynchusnerka), especially in the ethmoid region. The ethmoid region is found in the anterior region of the brain behind the ocular region. Unlike bacterial magnetite, the deposits found in the salmon brain do not contain any amorphous layer of Fe_3O_4 but is highly crystalline. The magnetite crystals have been found to be associated into chains that could be linear or in bundled formation. Each chain is found to consist of 13-45 magnetite crystals. The average crystal size was between 25-60 nm. All the magnetite crystals are found arranged with their (111) face perpendicular to the chain axis. All the magnetite crystals were found to possess cubo-octahedral geometry. What is amazing is that these crystals found in salmon do not exhibit any structural defects! The amount of magnetite in the salmon brain seems to increase with age.

Why does salmon produce magnetite? Well, the most likely explanation is because it uses these magnetite crystals for navigation with respect to the earth's magnetic field as a reference. This phenomenon is known as 'magnetoception' or 'magnetoreception'. The weak magnetic field of the earth can be detected through changes in the magnetic environment around these magnetite crystals present in the brain, which aids the fish to determine its coordinates with respect to the earth's magnetic field. Interestingly, magnetic material other than magnetite have also been found in other tissues such as the eye, skin and olfactory tissues of salmon, but these have no magnetoception role.

A similar mechanism for magnetoception has also been reported for several species of birds. One of the most widely investigated bird species for its astonishing navigation property is the 'homing pigeon'. Though the exact mechanism of navigation by these birds are yet to be agreed upon, the presence of considerable quantities of magnetite crystals in the beak and tissues of the pigeon seems to have a significant role in its navigation capabilities. Other suggestive navigation mechanisms apart from the magnetoception theory include light-dependent navigation and olfactory-based navigation. Honeybees and bats also have been demonstrated to exhibit magnetoception properties.

Did you know?

In 2009, a South African Information Technology Company sent a 4 GB flash drive containing data through a carrier pigeon named Winston from one of their offices to another office at Durban, 80 km away. At the same time, they also sent the same data to their Durban office through the broadband connection. The pigeon took sixty eight minutes to reach the Durban office and the office staff took about an hour to transfer the data from the flash drive to their systems. The total exercise took two hours six minutes and fifty seven seconds. On comparison, during the same duration, only 4% of the data was downloaded through the broadband connection!

Magnetite deposits have also been discovered in human tissues, especially in the brain. Though there remains some controversy regarding the concentration and localization of these magnetite deposits, it is now an accepted fact that human brain tissues and meninges (membrane layers that surround the brain and spinal cord) contain magnetite deposits. Another category of iron based deposits are ferritin moieties which is paramagnetic as compared to the ferromagnetic magnetite crystals. These magnetite deposits are prismatic as compared to the octahedral morphology commonly observed with geological samples indicating that these crystals have a different route of synthesis in vivo. The formation of these crystals having (111) crystal faces is most likely to resemble formation of a magnetosome in bacteria.

The biogenic magnetite crystalsin human brain have two different size ranges – 10-70 nm and a larger population between 90 and 200 nm. Also, it has been recently discovered that apart from the brain, tumour tissues also have magnetic crystal deposits. Hence it is unlikely that these could be used for magnetoceptive purposes. Then what is the role of such deposits in humans?One of the suggested roles indicates a potential gating molecule role for these crystals in mechanically sensitive ion channels. The opening and closing of ion channels in response to the extremely weak magnetic fields could change signal transduction, membrane polarity and cell response. The channel opening and closing could be triggered due to pressure changes accompanying change in orientation of the magnetic dipole in the crystals associated with the mechanically sensitive ion channel most probably through cytoskeletal proteins.

Another hypothesis is that the magnetite deposits may have a role in cell division and proliferation. However, more in-depth studies are required to probe the role of such biogenic magnetic crystals in humans. Interestingly, a gender-based difference in the amount of magnetite deposition has been also observed. Men show an increased amount of magnetite deposits when compared with women. The reason for this difference would come to light only after understanding the purpose of such deposits in humans!

Applications of Biogenic Nanoparticles

The biogenic nanoparticles have found wide range of applications in diverse fields. The silver and gold nanoparticles synthesized by biological systems exhibit excellent stability. Silver nanoparticles have been employed as anti-microbial agents in textiles, paper, domestic appliances, non-biofouling coatings, catheters, wound dressings etc. Their cytotoxicity is mainly attributed to their ability to alter membrane permeability and interfere with the DNA. They also could disrupt the electron transport components leading to cell death.The noble metal nanoparticles (silver and gold) also find applications as imaging agents. The presence of protein stabilizers around the nanoparticles enables functionalization of the nanoparticles with specific molecules that impart special characteristics to the nanoparticles. For example, attachment of a poly(ethylene glycol) chain confers long circulating characteristics to the particles while attachment of a target specific ligand enables selective internalization of these particles into specific cells. Figure represents a multi-functional magnetite nanoparticle.

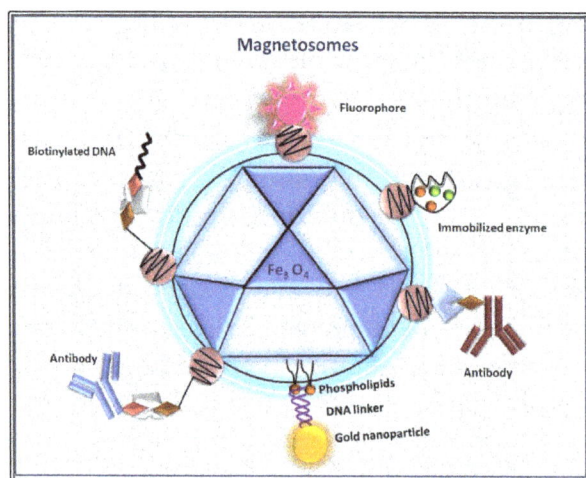

Multi-functional magnetite nanoparticle modified with a fluorescence moiety, antibodies for targeting, immobilized enzyme for catalysis and gold nanoparticle for photosensitivity and a DNA moiety

The semi-conductor nanoparticles find extensive applications in the electronics industries and for imaging applications. The magnetite nanoparticles have nearly monodisperse size distributions that are in the superparamagnetic regime. This property can be utilized inmagnetic separations, inducing hyperthermia in cancer cells as well as resonance imaging as contrast agents.

Stealth Nanoparticles

One of the major research areas in nanomedicine is in the development of nanocarriers to deliver drugs to specified locations. Though in vitro experiments employing the drug- loaded nanocarriers are successful, the efficiency of these drug carriers in vivo is not encouraging as they fail to reach their target. What could be wrong? There are numerous factors that can influence the efficiency of a drug carrierin vivo. These include the route of administration, preferential uptake by non-target cells (biodistribution) as well as the lifetime of the carrier within the biological system.There are many options for introduction of the nanocarrier into the biological system such as through nasal, transdermal, parenteral, oral, rectal or ocular routes. The efficiency of the system is different in each case. Why? The carrier molecule faces different types of barriers and filters in each case before it can enter the blood. As a result, the amount of drug-loaded carriers that enter into the blood circulation varies with the route of administration. Once it is in the blood, the carrier faces an entirely new challenge! Blood contains an arsenal of components that form part of the body's defense – namely the immune system. The nanocarrier is not native to the biological system and is considered a foreign body by the immune system.Hence, it is either degraded or eliminated from the biological systemas soon as the carrier is recognized by the immune system. As a result, the drug present in the nanocarrier never gets to reach its target! Therefore the carrier needs to escape recognition by the immune system components! The properties of the nanocarrier have a major influence on the response of the immune system as well as its interaction with other non-specific targets in the biological system. These include the particle size, surface charge, surface topography and hydrophilicity. As a result of such non-specific and undesirable interactions with non-specific targets as well as the immune system, the carrier and the drug loaded in the carrier do not get delivered to the target. Therefore, suitable modifications must be made to the nanocarrier to impart the ability to 'hide' from the immune system. The ability of a nanoparticle to evade immune recognition so as to enhance its circulation timein vivo and thereby its chances of reaching the target is known as 'stealth characteristic'.

One of the earliest attempts employed to evade the immune recognition was to flood the system with a placebo carrier before introducing the drug-loaded carrier. A placebo carrier does not contain any drug. The high concentrations of the placebo carrier will activate the immune components, which will then be engaged in eliminating these carriers. If the drug-loaded carrier is introduced at this time, the chances of it getting recognized and eliminated by the immune system is remote as the immune components are involved in destroying the placebo carrier. This diversionary strategy unfortunately comes with a price! The very high concentrations of the placebo carrier introduced into the system can cause undesirable side effects owing to the disruption of the immune system. Hence this strategy has been discarded.

Introduction of immune suppressive agents is another strategy that tends to disable the abilities of the immune system. Though this strategy is used in extreme cases, it is not desirable as the patient becomes vulnerable to other infections during this period. In recent years, a 'Trojan horse concept' has been employed to impart stealth characteristics to the nanocarrier to enable it to escape from immune recognition without causing any impairment to the immune system.

The origin of the Trojan horse concept…

In Greek mythology, a long war was fought between the kingdoms of Troy and Greece. The Greeks laid siege to the city of Troy for more than 10 years and still no end to the war was in sight. Hence, they decided to win the war by employing a stealth strategy. They constructed a large wooden horse with a secret compartment in which they hid some of their soldiers. Then they pretended to sail away making the Trojans think that the Greeks had given up. The unsuspecting Trojans took the horse inside their city to mark their victory. In the night, the soldiers inside the horse came out and opened the gates of the city to let in the Greek army, which then plundered the city and won the war! In the case of delivering the drugs to selected locations in the biological system, a similar strategy has to be employed to trick the immune system. The carrier becomes the Trojan horse carrying the 'soldiers', namely the drug molecules. The immune system fails to recognize the danger posed and once the carrier enters the 'city of Troy', which in this case is the target cell, the drug molecules come out of the carrier and destroy the cell. Hence the Trojan horse concept is very relevant in nanomedicine to design stealth carrier systems!

In order to devise strategies to impart stealth characteristics to the carrier, one has to understand the pathways leading to immune recognition of foreign bodies. The response of the blood components to the nanocarrier will be of prime importance in deciding the fate of the nanocarrier in the biological system. Apart from the nature of the nanocarrier, the flow parameters of the blood also influence the interactions between the blood and the carrier. These include the kind of stress experienced by the carrier surface owing to the flow of blood – static, laminar or vortex, and the ratio of the blood volume in contact with the nanocarrier surface. Any carrier that does not elicit much response from the blood components is termed as 'hemocompatible'. The search for the ideal hemocompatible material is still on. Apart from the natural vascular endothelial cells lining the inner side of the blood vessel, no material can be classified as 100% hemocompatible!

Nanocarriers

A nanocarrier is nanomaterial being used as a transport module for another substance, such as a drug. Commonly used nanocarriers include micelles, polymers, carbon-based materials, liposomes and other substances. Nanocarriers are currently being studied for their use in drug delivery and their unique characteristics demonstrate potential use in chemotherapy.

Characterization

Nanocarriers range from sizes of diameter 1–1000 nm, however due to the width of microcapillaries being 200 nm, nanomedicine often refers to devices <200 nm. Because of their small size, nanocarriers can deliver drugs to otherwise inaccessible sites around the body. Since nanocarriers are so small, it is oftentimes difficult to provide large drug doses using them. The emulsion techniques used to make nanocarriers also often result in low drug loading and drug encapsulation, providing a difficulty for the clinical use.

Types

Nanocarriers discovered thus far include polymer conjugates, polymeric nanoparticles, lipid-based carriers, dendrimers, carbon nanotubes, and gold nanoparticles. Lipid-based carriers include both liposomes and micelles. Examples of gold nanoparticles are gold nanoshells and nanocages. Different types of nanomaterial being used in nanocarriers allows for hydrophobic and hydrophilic drugs to be delivered throughout the body. Since the human body contains mostly water, the ability to deliver hydrophobic drugs effectively in humans is a major therapeutic benefit of nanocarriers. Micelles are able to contain either hydrophilic or hydrophobic drugs depending on the orientation of the phospholipid molecules. Some nanocarriers contain nanotube arrays allowing them to contain both hydrophobic and hydrophilic drugs.

One potential problem with nanocarriers is unwanted toxicity from the type of nanomaterial being used. Inorganic nanomaterial can also be toxic to the human body if it accumulates in certain cell organelles. New research is being conducted to invent more effective, safer nanocarriers. Protein based nanocarriers show promise for use therapeutically since they occur naturally, and generally demonstrate less cytotoxicity than synthetic molecules.

Targeted Drug Delivery

Nanocarriers are useful in the drug delivery process because they can deliver drugs to site-specific targets, allowing drugs to be delivered in certain organs or cells but not in others. Site-specificity is a major therapeutic benefit since it prevents drugs from being delivered to the wrong places. Nanocarriers show promise for use in chemotherapy because they can help decrease the adverse, broader-scale toxicity of chemotherapy on healthy, fast growing cells around the body. Since chemotherapy drugs can be extremely toxic to human cells, it is important that they are delivered to the tumor without being released into other parts of the body. Four methods in which nanocarriers can deliver drugs include passive targeting, active targeting, pH specificity, and temperature specificity.

Passive Targeting

Passive targeting refers to a nanocarrier's ability to travel down a tumor's vascular system, become trapped, and accumulate in the tumor. This accumulation is caused by the enhanced permeability and retention effect which refers to the poly(ethylene oxide) (PEO) coating on the outside of many nanocarriers. PEO allows nanocarriers to travel through the leaky vasculature of a tumor, where they are unable to escape. The leaky vasculature of a tumor is the network of blood vessels that form in a tumor, which contain many small pores. These pores allow nanocarriers in, but also contain many bends that allow the nanocarriers to become trapped. As more nanocarriers become trapped, the drug accumulates at the tumor site. This accumulation cause large doses of the drug to be delivered directly to the tumor site. PEO may also have some adverse effects on cell-nanocarrier interactions, weakening the effects of the drug, since many nanocarriers must be incorporated into the cells before the drugs can be released.

Active Targeting

Active targeting involves the incorporation of targeting modules such as ligands or antibodies on the surface of nanocarriers that are specific to certain types of cells around the body. Nanocarriers have such a high surface-area to volume ratio allowing for multiple ligands to be incorporated on their surfaces. These targeting modules allow for the nanocarriers to be incorporated directly inside of cells, but also have some drawbacks. Ligands may cause nanocarriers to become slightly more toxic due to non-specific binding, and positive charges on ligands may decrease drug delivery efficiency once inside of cells. Active targeting has been shown to help overcome multi-drug resistance in tumor cells.

pH specificity

Certain nanocarriers will only release the drugs they contain in specific pH ranges. pH specificity also allows nanocarriers to deliver drugs directly to a tumor site. Tumors are generally more acidic than normal human cells, with a pH around 6.8. Normal tissue has a pH of around 7.4. Nanocarriers that only release drugs at certain pH ranges can therefore be used to release the drug only within acidic tumor environments. High acidic environments cause the drug to be released due to the acidic environment degrading the structure of the nanocarrier. These nanocarriers will not release drugs in neutral or basic environments, effectively targeting the acidic environments of tumors while leaving normal body cells untouched. This pH sensitivity can also be induced in micelle systems by adding copolymer chains to micelles that have been determined to act in a pH independent manor. These micelle-polymer complexes also help to prevent cancer cells from developing multi-drug resistance. The low pH environment triggers a quick release of the micelle polymers, causing a majority of the drug to be released at once, rather than gradually like other drug treatments. This quick release mechanism significantly decreases the time it takes for anticancer drugs to kill a tumor, effectively preventing the tumor from having time to undergo mutations that would render it drug resistant.

Temperature Specificity

Some nanocarriers have also been shown to deliver drugs more effectively at certain temperatures. Since tumor temperatures are generally higher than temperatures throughout the rest of the body, around 40 °C, this temperature gradient helps act as safeguard for tumor-specific site delivery.

Uses

Most of research on nanocarriers is being applied to their potential use in drug delivery, especially in chemotherapy. Since nanocarriers can be used to specifically target the small pores, lower pH's, and higher temperatures of tumors, they have the potential to lower the toxicity of many chemotherapy drugs. Also, since almost 75% of anticancer drugs are hydrophobic, and therefore demonstrate difficulty in delivery inside human cells, the use of micelles to stabilize, and effectively mask the hydrophobic nature of hydrophobic drugs provides new possibilities for hydrophobic anticancer drugs.

In vitro

In vitro studies are performed with microorganisms, cells, or biological molecules outside their normal biological context. Colloquially called "test-tube experiments", these studies in biology and its subdisciplines are traditionally done in labware such as test tubes, flasks, Petri dishes, and microtiter plates. Studies conducted using components of an organism that have been isolated from their usual biological surroundings permit a more detailed or more convenient analysis than can be done with whole organisms; however, results obtained from *in vitro* experiments may not fully or accurately predict the effects on a whole organism. In contrast to *in vitro* experiments, *in vivo* studies are those conducted in animals, including humans, and whole plants.

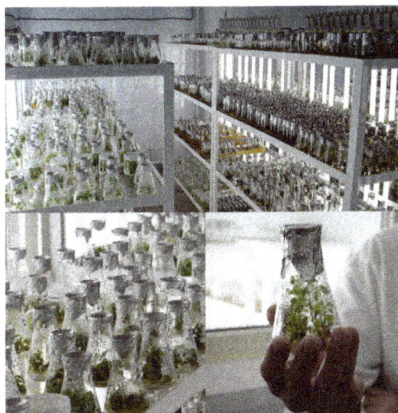

Cloned plants in vitro

Definition

In vitro studies are conducted using components of an organism that have been isolated from their usual biological surroundings, such as microorganisms, cells, or biological molecules. For example, microrganisms or cells can be studied in artificial culture media, and proteins can be examined in solutions. Colloquially called "test-tube experiments", these studies in biology, medicine, and their subdisciplines are traditionally done in test tubes, flasks, Petri dishes, etc. They now involve the full range of techniques used in molecular biology, such as the omics.

In contrast, studies conducted in living beings (microorganisms, animals, humans, or whole plants) are called *in vivo* .

Examples

Examples of *in vitro* studies include: the isolation, growth and identification of cells derived from multicellular organisms in (cell or tissue culture); subcellular components (e.g. mitochondria or ribosomes); cellular or subcellular extracts (e.g. wheat germ or reticulocyte extracts); purified molecules such as proteins, DNA, or RNA); and the commercial production of antibiotics and other pharmaceutical products. Viruses, which only replicate in living cells, are studied in the laboratory in cell or tissue culture, and many animal virologists refer to such work as being *in vitro* to distinguish it from *in vivo* work in whole animals.

- Polymerase chain reaction is a method for selective replication of specific DNA and RNA sequences in the test tube.

- Protein purification involves the isolation of a specific protein of interest from a complex mixture of proteins, often obtained from homogenized cells or tissues.

- *In vitro* fertilization is used to allow spermatozoa to fertilize eggs in a culture dish before implanting the resulting embryo or embryos into the uterus of the prospective mother.

- *In vitro* diagnostics refers to a wide range of medical and veterinary laboratory tests that are used to diagnose diseases and monitor the clinical status of patients using samples of blood, cells, or other tissues obtained from a patient.

- *In vitro* testing has been used to characterize specific adsorption, distribution, metabolism, and excretion processes of drugs or general chemicals inside a living organism; for example, Caco-2 cell experiments can be performed to estimate the absorption of compounds through the lining of the gastrointestinal tract; The partitioning of the compounds between organs can be determined to study distribution mechanisms; Suspension or plated cultures of primary hepatocytes or hepatocyte-like cell lines (HepG2, HepaRG) can be used to study and quantify metabolism of chemicals. These ADME process parameters can then be integrated into so called "physiologically based pharmacokinetic models" or PBPK.

Advantages

In vitro studies permit a species-specific, simpler, more convenient, and more detailed analysis than can be done with the whole organism. Just as studies in whole animals more and more replace human trials, so are *in vitro* studies replacing studies in whole animals.

Simplicity

Living organisms are extremely complex functional systems that are made up of, at a minimum, many tens of thousands of genes, protein molecules, RNA molecules, small organic compounds, inorganic ions, and complexes in an environment that is spatially organized by membranes, and in the case of multicellular organisms, organ systems. These myriad components interact with each other and with their environment in a way that processes food, removes waste, moves components to the correct location, and is responsive to signalling molecules, other organisms, light, sound, heat, taste, touch, and balance.

Top view of a Vitrocell mammalian exposure module "smoking robot", (lid removed) view of four separated wells for cell culture inserts to be exposed to tobacco smoke or an aerosol for an in vitro study of the effects

This complexity makes it difficult to identify the interactions between individual components and to explore their basic biological functions. *In vitro* work simplifies the system under study, so the investigator can focus on a small number of components.

For example, the identity of proteins of the immune system (e.g. antibodies), and the mechanism by which they recognize and bind to foreign antigens would remain very obscure if not for the extensive use of *in vitro* work to isolate the proteins, identify the cells and genes that produce them, study the physical properties of their interaction with antigens, and identify how those interactions lead to cellular signals that activate other components of the immune system.

Species Specificity

Another advantage of *in vitro* methods is that human cells can be studied without "extrapolation" from an experimental animal's cellular response.

Convenience, Automation

In vitro methods can be miniaturized and automated, yielding high-throughput screening methods for testing molecules in pharmacology or toxicology.

Disadvantages

The primary disadvantage of *in vitro* experimental studies is that it may be challenging to extrapolate from the results of *in vitro* work back to the biology of the intact organism. Investigators doing *in vitro* work must be careful to avoid over-interpretation of their results, which can lead to erroneous conclusions about organismal and systems biology.

For example, scientists developing a new viral drug to treat an infection with a pathogenic virus (e.g. HIV-1) may find that a candidate drug functions to prevent viral replication in an *in vitro* setting (typically cell culture). However, before this drug is used in the clinic, it must progress through a series of *in vivo* trials to determine if it is safe and effective in intact organisms (typically small animals, primates, and humans in succession). Typically, most candidate drugs that are effective *in vitro* prove to be ineffective *in vivo* because of issues associated with delivery of the drug to the affected tissues, toxicity towards essential parts of the organism that were not represented in the initial *in vitro* studies, or other issues.

In Vitro to *in Vivo* Extrapolation

Results obtained from *in vitro* experiments cannot usually be transposed, as is, to predict the reaction of an entire organism *in vivo*. Building a consistent and reliable extrapolation procedure from *in vitro* results to *in vivo* is therefore extremely important. Solutions include:

- Increasing the complexity of *in vitro* systems to reproduce tissues and interactions between them (as in "human on chip" systems)

- Using mathematical modeling to numerically simulate the behavior of the complex system, where the *in vitro* data provide model parameter values

These two approaches are not incompatible; better *in vitro* systems provide better data to mathematical models. However, increasingly sophisticated *in vitro* experiments collect increasingly numerous, complex, and challenging data to integrate. Mathematical models, such as systems biology models, are much needed here.

Extrapolating in Pharmacology

In pharmacology, IVIVE can be used to approximate pharmacokinetics (PK) or pharmacodynamics (PD). Since the timing and intensity of effects on a given target depend on the concentration time course of candidate drug (parent molecule or metabolites) at that target site, *in vivo* tissue and organ sensitivities can be completely different or even inverse of those observed on cells cultured and exposed *in vitro*. That indicates that extrapolating effects observed *in vitro* needs a quantitative model of *in vivo* PK. Physiologically based PK (PBPK) models are generally accepted to be central to the extrapolations.

In the case of early effects or those without intercellular communications, the same cellular exposure concentration is assumed to cause the same effects, both qualitatively and quantitatively, *in vitro* and *in vivo*. In these conditions, developing a simple PD model of the dose–response relationship observed *in vitro*, and transposing it without changes to predict *in vivo* effects is not enough.

Immune Recognition Pathways

One of the first events that occur when a nanocarrierenters the blood stream is 'opsonization'. Blood consists of a huge number of proteins and ions. When the nanocarrier comes into contact with these components, adsorption of the blood proteins on the surface of the nanocarrier occurs. This is a rapid process and is diffusion- controlled in the initial phases. This means that the first interacting components will be the hydrated ions followed by the more mobile proteins. Among the ions, it has been found that the divalent calcium and magnesium ions are the most active. The later stages of opsonization are affinity-controlled as the larger proteins that have greater affinity to the nanocarrier surface displace the weakly adsorbed species. The protein adsorption process over the nanocarrier is thus a competitive process and the sequence of proteins that adsorb on the surface and then on each other is described by 'Vroman effect'. The major blood proteins involved in the adsorption process are albumin, globulin, fibrinogen, fibronectin, Factor XII and high molecular weight kininogen. These proteins are referred to as the 'Opsonins'.

What is the consequence of this protein adsorption? Well, the opsonin proteins can interact with the receptors in the cells of the mononuclear phagocytic system (MPS) that form part of the cell-mediated immune response. This interaction leads to the activation of the monocytes and macrophages,which results in the uptake the nanocarrier by the macrophages or activationof the other components of the immune system. The nanocarrier is taken up by the macrophage by forming a vesicle termed as phagosome. This then fuses with the lysosome containing an array of hydrolytic and degradative enzymes that destroys the nanocarrier. This process of uptake of the carrier by the macrophages or other components of the MPS is termed as 'phagocytosis'.

Why is this process termed 'opsonization'? The term 'opsonin' is derived from the Greek language and means 'seasoning' or 'sauce'. Similar to the process of seasoning food items with herbs, spices or other ingredients to improve its flavor and edibility, the nanoparticle surface is coated or 'seasoned' with proteins to improve its uptake and digestion by the cells of the MPS! Hence this sequence of events is referred to as opsonization. Figure shows a cartoon depicting the sequence of events occurring during opsonization.

Why should opsonization drive phagocytosis? Can't the macrophage phagocytose a foreign body without opsonization? The immune system has evolved mainly to combat and eliminate pathogens. Most pathogens have an outer membrane that is predominantly negatively charged. Unfortunately, the macrophages and the other cells of the MPS also possess a negatively charged cell membrane. Hence uptake of the pathogen by the macrophages is retarded due to the existence of greater electrostatic repulsive forces. Hence, the immune system has evolved an effective strategy of coating the surface of the pathogen with the opsonin proteins that will mask the surface charges of the pathogen thereby facilitating uptake and subsequent degradation by the macrophages.Though it is not necessary that all nanoparticle surfaces will be negatively charged or even charged at all, the process of opsonization is the first they will encounter when they enter the blood stream.

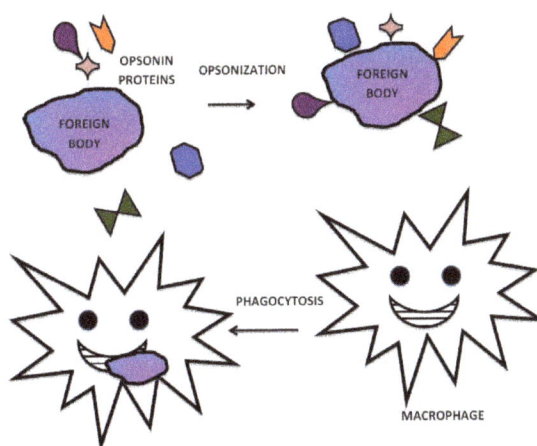

Cartoon showing the sequence of events during opsonisation

What will happen if the nanocarrier could not be degraded by the MPS? In this event,the entire carrier molecule or part of it is anchored on the cell surface and 'presented' or exhibitedto the external environment by the MPS cells using major histocompatibility (MHC) proteins. This results in the activation of another component of the immune system, namely the T-lymphocytes, which in turn activate the B-lymphocytes leading to production of antibodies against the foreign body.

Among the range of opsonins that can adsorb on to a foreign body surface are a group of proteins called as the complement proteins. If the complement proteins adsorb on to the surface of the nanocarrier, then another cascade of events leading to the activation of the complement system is also triggered.

Opsonin

An opsonin is any molecule that enhances phagocytosis by marking an antigen for an immune response or marking dead cells for recycling (i.e., causes the phagocyte to "relish" the marked cell). *Opson* in ancient Greece referred to the delicious side-dish of any meal, versus the *sitos*, or the staple of the meal.

Opsonization (also, opsonisation) is the molecular mechanism whereby molecules, microbes, or apoptotic cells are chemically modified to have stronger interactions with - to be more "delicious" to - cell surface receptors on phagocytes and NK cells. With the antigen coated in opsonins, binding to immune cells is greatly enhanced. Opsonization also mediates phagocytosis via signal cascades from cell surface receptors.

Opsonins aid the immune system in a number of ways. In a healthy individual, they mark dead and dying self cells for clearance by macrophages and neutrophils, activate complement proteins, and target cells for destruction through the action of natural killer (NK) cells.

Mechanism

All cell membranes have negative charges (Zeta potential) which makes it difficult for two cells to come close together. When opsonins bind to their targets they boost the kinetics of phagocytosis by favoring interaction between the opsonin and cell surface receptors on immune cells. This overrides the negative charges from cell membranes. This principle holds true for clearance of pathogens as well as dead or dying *self* cells.

Varieties

Different opsonins perform different functions. Opsonin molecules include:

Antibodies

Antibodies are part of the adaptive immune response and are generated by B cells in response to antigen exposure. The Fab region of the antibody binds to the antigen, whereas the Fc region of the antibody binds to an Fc receptor on the phagocyte, facilitating phagocytosis. The antigen-antibody complex can also activate complement through the classical complement pathway. Phagocytic cells do not have an Fc receptor for immunoglobulin M (IgM), making IgM ineffective in assisting phagocytosis alone. However, IgM is extremely efficient at activating complement and is, therefore, considered an opsonin. IgG antibodies are also capable of binding immune effector cells via their Fc domain, triggering a release of lysis products from the bound immune effector cell (monocytes, neutrophils, eosinophils, and natural killer cells). This process, called antibody-dependent cellular cytotoxicity, can cause inflammation of surrounding tissues and damage to healthy cells.

Complement Proteins

The complement system is a part of the innate immune response. C3b, C4b, and C1q are important complement molecules that serve as opsonins. As a part of the alternative complement pathway, the spontaneous activation of a complement cascade converts C3 to C3b, a component that can serve as an opsonin when bound to an antigen's surface. Antibodies can also activate complement via the classical pathway, resulting in deposition of C3b and C4b onto the antigen surface. After C3b has bound to the surface of an antigen, it can be recognized by phagocyte receptors that signal for phagocytosis. Complement receptor 1 is expressed on all phagocytes and recognizes a number of complement opsonins, including C3b and C4b which are both parts of C3-convertase. C1q, a member of the C1 complex, is able to interact with the Fc region of antibodies.

Circulating Proteins

Pentraxins, collectins, and ficolins are all circulating proteins that are capable of serving as opsonins. They are secreted Pattern recognition receptors (PRRs). These molecules coat the microbes as opsonins and enhance neutrophil reactivity against them through a number of mechanisms.

Targets

Apoptotic Cells

Apoptosis is related to low tissue inflammation. A number of opsonins play a role in marking apoptotic cells for phagocytosis without a pro-inflammatory response.

Members of the pentraxin family can bind to apoptotic cell membrane components like phosphatidylcholine (PC) and phosphatidylethanolamine (PE). IgM antibodies also bind to PC. Collectin molecules such as mannose-binding lectin (MBL), surfactant protein A (SP-A), and SP-D interact with unknown ligands on apoptotic cell membranes. When bound to the appropriate ligand these molecules interact with phagocyte receptors, enhancing phagocytosis of the marked cell.

C1q is capable of binding directly to apoptotic cells. It can also indirectly bind to apoptotic cells via intermediates like IgM autoantibodies, MBL, and pentraxins. In both cases C1q activates complement, resulting in the cells being marked for phagocytosis by C3b and C4b. C1q is an important contributor to the clearance of apoptotic cells and debris. This process usually occurs in late apoptotic cells.

Opsonization of apoptotic cells occurs by different mechanisms in a tissue-dependent pattern. For example, while C1q is necessary for proper apoptotic cell clearance in the peritoneal cavity, it is not important in the lungs where SP-D plays an important role.

Pathogens

As part of the late stage adaptive immune response, pathogens and other particles are marked by IgG antibodies. These antibodies interact with Fc receptors on macrophages and neutrophils resulting in phagocytosis. The C1 complement complex can also interact with the Fc region of IgG and IgM immune complexes activating the classical complement pathway and marking the antigen with C3b. C3b can spontaneously bind to pathogen surfaces through the alternative complement pathway. Furthermore, pentraxins can directly bind to C1q from the C1 complex.

SP-A opsonizes a number of bacterial and viral pathogens for clearance by lung alveolar macrophages.

Phagocytosis

In cell biology, phagocytosis is the process by which a cell—often a phagocyte or a protist—engulfs a solid particle to form an internal compartment known as a phagosome. It is distinct from other forms of endocytosis like pinocytosis that involves the internalization of extracellular liquids. Phagocytosis is involved in the acquisition of nutrients for some cells. The process is homologous to eating at the level of single-celled organisms; in multicellular animals, the process has been adapted to eliminate debris and pathogens, as opposed to taking in fuel for cellular processes, except in the case of the animal *Trichoplax*.

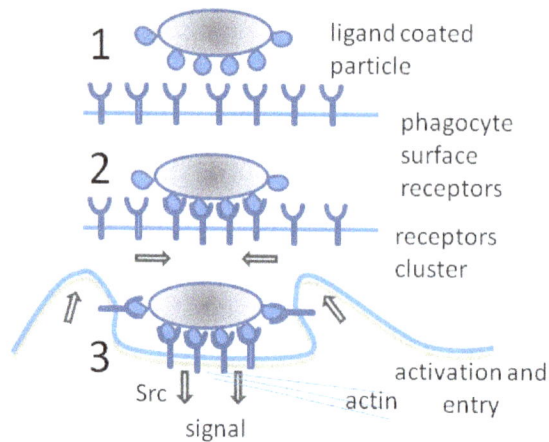

Phagocytosis in three steps:
1. Unbound phagocyte surface receptors do not trigger phagocytosis.
2. Binding of receptors causes them to cluster.
3. Phagocytosis is triggered and the particle is taken up by the phagocyte.

In an organism's immune system, phagocytosis is a major mechanism used to remove pathogens and cell debris. For example, when a macrophage ingests a pathogenic microorganism, the pathogen becomes trapped in a phagosome which then fuses with a lysosome to form a phagolysosome. Within the phagolysosome, enzymes and toxic peroxides digest the pathogen. Bacteria, dead tissue cells, and small mineral particles are all examples of objects that may be phagocytized. The process has triggered the name "Phagocytes" for the 1st line of defence in the immune system.

History

Phagocytosis was first noted by Canadian physician William Osler (1876), and later studied and named by Élie Metchnikoff (1880, 1883).

In Immune System

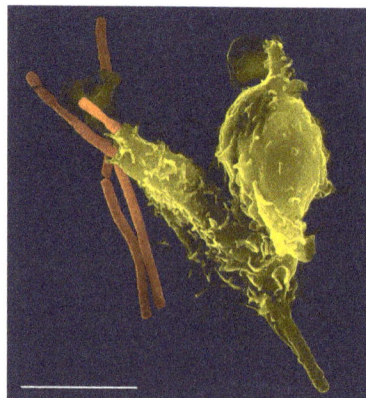

Scanning electron micrograph of a phagocyte (yellow, right) phagocytosing anthrax bacilli (orange, left)

Phagocytosis in mammalian immune cells is activated by attachment to pathogen-associated molecular patterns (PAMPS), which leads to NF-κB activation. Opsonins such as C3b and antibodies can act as attachment sites and aid phagocytosis of pathogens.

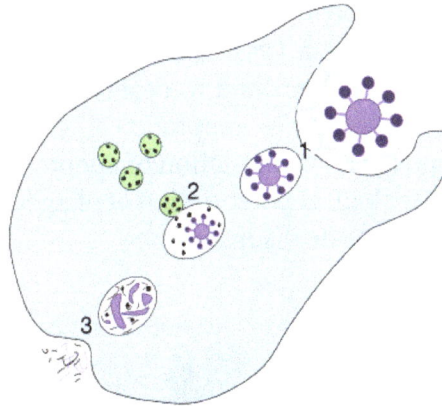

Phagocytosis is the process in which a cell engulfs a particle,
digests it, and expels the waste products.

Process of phagocytosis:

1. A particle is ingested by a phagocyte after antigens are recognized which results in the formation of a phagosome.

2. The fusion of lysosomes with the phagosome creates a phagolysosome.

3. The particle is broken down by the digestive enzymes found in the lysosomes. The resulting waste material is discharged from the phagocyte by exocytosis.

Engulfment of material is facilitated by the actin-myosin contractile system. The phagosome of ingested material is then fused with the lysosome, forming a phagolysosome and leading to degradation.

Degradation can be oxygen-dependent or oxygen-independent.

- Oxygen-dependent degradation depends on NADPH and the production of reactive oxygen species. Hydrogen peroxide and myeloperoxidase activate a halogenating system, which leads to the creation of hypochlorite and the destruction of bacteria.

- Oxygen-independent degradation depends on the release of granules, containing proteolytic enzymes such as defensins, lysozyme, and cationic proteins. Other antimicrobial peptides are present in these granules, including lactoferrin, which sequesters iron to provide unfavourable growth conditions for bacteria.

It is possible for cells other than dedicated phagocytes (such as dendritic cells) to engage in phagocytosis. Some white blood cells in human immune system perform phagocytosis by gulping in some pathogenic and disease causing cells.

In Apoptosis

Following apoptosis, the dying cells need to be taken up into the surrounding tissues by macrophages in a process called efferocytosis. One of the features of an apoptotic cell is the presentation of a variety of intracellular molecules on the cell surface, such as calreticulin, phosphatidylserine (From the inner layer of the plasma membrane), annexin A1, oxidised LDL and altered glycans.

These molecules are recognised by receptors on the cell surface of the macrophage such as the phosphatidylserine receptor or by soluble (free-floating) receptors such as thrombospondin 1, GAS6, and MFGE8, which themselves then bind to other receptors on the macrophage such as CD36 and alpha-v beta-3 integrin. Defects in apoptotic cell clearance is usually associated with impaired phagocytosis of macrophages. Accumulation of apoptotic cell remnants often causes autoimmune disorders; thus pharmacological potentiation of phagocytosis has a medical potential in treatment of certain forms of autoimmune disorders.

In Protists

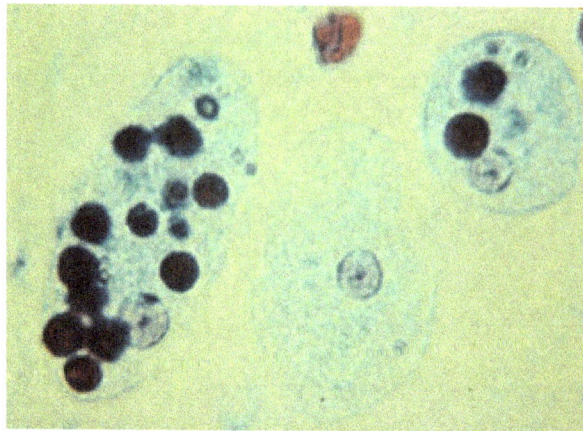

Trophozoites of *Entamoeba histolytica* with ingested erythrocytes

In many protists, phagocytosis is used as a means of feeding, providing part or all of their nourishment. This is called phagotrophic nutrition, distinguished from osmotrophic nutrition which takes place by absorption.

- In some, such as amoeba, phagocytosis takes place by surrounding the target object with pseudopods, as in animal phagocytes. In humans, *entamoeba histolytica* can phagocytose red blood cells. Entamoeba histolytica is an anaerobic parasitic protozoan, part of the genus Entamoeba. Predominantly infecting humans and other primates, E. histolytica is estimated to infect about 50 million people worldwide. Previously, it was thought that 10% of the world population was infected, but these figures predate the recognition that at least 90% of these infections were due to a second species, E. dispar. Mammals such as dogs and cats can become infected transiently, but are not thought to contribute significantly to transmission.The word histolytic literally means "Tissue destroyer". This process is known as "erythrophagocytosis", and is considered the only reliable way to distinguish *Entamoeba histolytica* from noninvasive species such as *Entamoeba dispar*.

- Ciliates also engage in phagocytosis. In ciliates there is a specialized groove or chamber in the cell where phagocytosis takes place, called the cytostome or mouth.

As in phagocytic immune cells, the resulting phagosome may be merged with lysosomes containing digestive enzymes, forming a phagolysosome. The food particles will then be digested, and the released nutrients are diffused or transported into the cytosol for use in other metabolic processes.

Mixotrophy can involve phagotrophic nutrition and phototrophic nutrition.

Complement Activation

Complement is referred to a group of proteins that form part of the innate immune system, which does not change over the course of an individual's lifetime due to environmental factors. The complement proteins are synthesized in the liver. Some of the complement proteins possess enzymatic activity but under normal conditions, they exist as zymogens. Zymogens require a trigger such as lysis of a polypeptide sequence to transform into an active enzyme. The introduction of a foreign body can trigger this transformation resulting in a cascade of events leading to enhanced phagocytosis, recruitment of other immune components and formation of a membrane attack complex (MAC) to destruct the foreign body. The activation of the complement system can occur in three different modes – classical pathway, alternative pathway and lectin pathway. All three pathways converge to a common pathway that results in formation of the membrane attack complex that alters the membrane permeability of a pathogen resulting in its death.

The classical pathway for complement activation is initiated by the formation of an immune complex (antigen-antibody complex) involving the immunoglobulins IgM or IgG (antibodies). The immune complex binds to the complement protein complex C1. The C1 complex comprises one C1q, two C1r and two C1s. If the immune complex binds to C1q or if the C1q directly binds to the surface of the foreign body, a conformational change occurs in the C1 complex. This change activates the C1r sub-units. The C1r is a serine protease, i.e., it is an enzyme containing serine in its active site, and can cleave peptide linkages. The substrate of C1r is the C1s sub-unit, which on lysis gets activated. The C1s also is a serine protease. The C1 complex containing the active sub-units of C1r and C1s now act upon two other complement proteins C4 and C2 to form C2a, C2b, C4a and C4b fragments. The lysis of the proteins does not form symmetrical fragments (i.e., same sized fragments). The smaller fragment is denoted by the suffix 'a' and the larger fragment is denoted by the suffix 'b'. The C4b and C2a fragments bind to form another enzyme known as the C3 convertase.

The C4b2a acts on the complement protein C3 to form C3a and C3b. The C3b fragment binds with the C42b complex to from yet another new enzyme C4b2a3b known as the C5 convertase. The C5 convertase cleaves the complement protein C5 to form C5a and C5b. The C5b then associates with the complement proteins C6, C7, C8 and multiple C9 in a sequential manner to form the membrane attack complex (MAC). The MAC inserts into the microbial membrane to form a pore leading to drastic changes in the permeability resulting in the cell death.

What is the fate of the smaller fragments C4a, C3 and C5a that are formed during the complement activation? These are anaphylatoxins that promote inflammatory reactions in the body. They stimulate the cells of the mononuclear phagocyte system (MPS) and the mast cells (cells with granules containing histamine, heparin, serine proteases, leukotrienes, prostaglandins, cytokines etc.; it is found in many tissues and is thought to play a key role in wound healing) to degranulate and release their contents that contain many lytic enzymes as well as chemokines and cytokines. The chemokines serve to attract the other components of the immune system from remote locations in the body to the site of release leading to an increase in the immune activity at the site of release. This migration of the immune components occurs towards increasing concentration gradientand is referred to as 'chemotaxis'. Thus, the formation of the anaphylatoxins during the complement activation leads to an amplification of the immune response against the foreign body. It is interesting to note that the activation of one type of immune response also leads to the activation of other classes of immune response in the biological system.

In the case of the alternative pathway, the activation involves the binding of the C3 protein with the foreign body. The C3 protein under normal conditions undergoes spontaneous lysis to form a nascent active form denoted as C3b-like-C3 or C3b*. This fragment is unstable and is inactivated by hydrolysis under normal conditions. The C3b- like-C3 contains a thioester group using which it can bind to surfaces containing nucleophilic groups (containing electron rich centres) such as amino group (-NH$_2$) or hydroxyl group (-OH). Such binding is prevented in the normal cells due to the presence of inhibitory factors, namely Factor H and Factor I, on the cell surfaces. Such factors are however lacking in the foreign body and hence the binding of the C3b-like-C3 can occur on the foreign body. In the presence of a foreign body, the C3b-like-C3 binds to the surface by cleavage of the thioester link to form a covalent ester link with the surface of the foreign body. In the absence of reactive functional groups on the surface of the foreign body, this binding occurs through the opsonin proteins adsorbed on the surface of the foreign body. Figure shows the reaction that occurs when C3b-like-C3 encounters a foreign body.

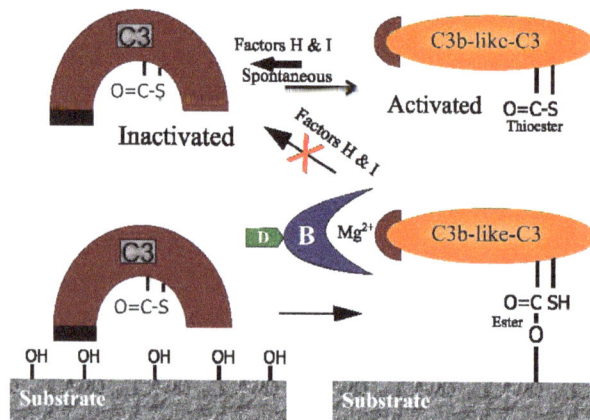

Binding of C3b-like-C3 to the surface of a foreign body

Once bound to the surface, the Factors H and I cannot inhibit the C3b-like-C3 anymore. In the presence of divalent magnesium ions, the C3b-like-Cr in the presence of Factor D, cleavesthe protein Factor B to Ba and Bb fragments The C3b-like-C3 associates with the larger Bb fragment to form the C3bBb complex. This complex is stabilized by the protein properidin (also known as Factor P). The stabilized C3bBb complex is now called the C3 convertase. It cleaves C3 to form C3a and C3b, which then associates with C3bBb to from the C5 convertase C3bBb3b. The C5 convertase cleaves C5 to C5a and C5b. The C5b associates with C6, C7, C8 and multiple C9 proteins to form the MAC.

The lectin pathway is initiated by the binding of the mannose binding lectin (MBL) protein to the mannose moiety found in the surface glycoproteins of microorganisms. This binding brings about the activation of MBL-associated serine proteases 1 and 2 (MASP-1 and MASP-2), which then cleave the complement proteins C4 and C2 into C4a, C4b, C2a and C2b fragments. The C4b and C2a bind to form C4b2a complex that functions as the C3 convertase. The rest of the sequence is common as in the classical and alternative pathways. Figure summarizes the events leading to activation of complement through different routes.

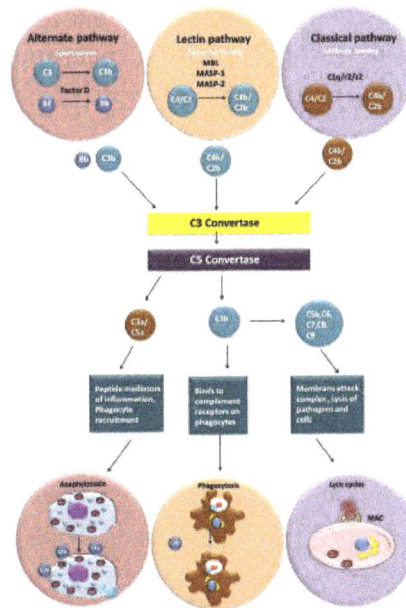

Summary of the events leading to complement activation through different pathways

It is evident that regardless of the route of activation, the activation of the complement protein C_3 remains the vital step that can lead to enhanced phagocytosis and destruction of the foreign body. The most common pathway that gets activated in the presence of a nanocarrier is the alternative pathway. The high surface-to-volume ratio of the nanocarrier enables more effective adsorption of the opsonin proteins and complement leading to rapid activation of the complement as well as phagocytosis. In the case of non- membranous carriers, the formation of MAC as a result of complement activation might not be effective in destruction of the carrier. But the generation of the anaphylatoxins and cytokines can aid in the recruitment of the MPS cells and the T-lymphocytes that can effectively destroy the carrier. Thus, the complement by itself is not effective in eliminating all foreign bodies, but it serves to alert and activate other powerful immune components that can eliminate the threat. Therefore suppression of opsonization might offer a good option to retard the complement activation and hence the immune recognition and response.

Coagulation Pathway

The blood also contains components of the coagulation pathway. What factors activate the coagulation pathway? The activation of the coagulation cascade could occur either due to intrinsic (direct contact pathway) or extrinsic(tissue contact pathway) pathways.The extrinsic pathway is activated with the release of thrombogenic tissue factor from the tissues as well as exposure of collagen from the extracellular matrix of injured tissue. The activation of the intrinsic pathway requires activation of many proteins in a sequential manner. Co-factors such as vitamin K and calcium ions are essential for all the coagulation processes. Figure depicts the various factors involved in the activation of the coagulation cascade by different types of nanoparticles.

Chemotaxis Assay

Chemotaxis assays are experimental tools for evaluation of chemotactic ability of prokaryotic or eukaryotic cells. A wide variety of techniques have been developed. Some techniques are qualitative - allowing an investigator to approximately determine a cell's chemotacic affinity for an analyte - while others are quantitative, allowing a precise measurement of this affinity.

Quality Control

In general, the most important requisite is to calibrate the incubation time of the assay both to the model cell and the ligand to be evaluated. Too short incubation time results no cells in the sample, while too long time perturbs the concentration gradients and measures more chemokinetic than chemotactic responses.

The most commonly used techniques are grouped into two main groups:

Agar-plate Techniques

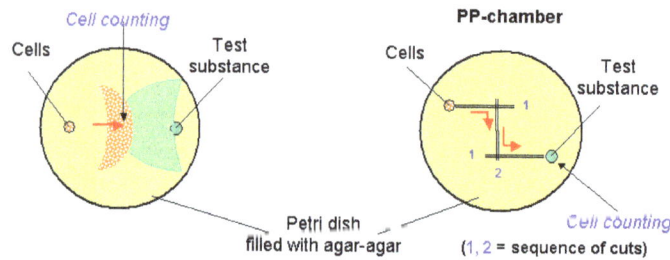

Chemotaxis assays with agar plates

This way of evaluation deals with agar-agar or gelatine containing semi-solid layers made prior to the experiment. Small wells are cut into the layer and filled with cells and the test substance. Cells can migrate towards the chemical gradient in the semi solid layer or under the layer as well. Some variations of the technique deal also with wells and parallel channels connected by a cut at the start of the experiment (PP-technique). Radial arrangement of PP-technique (3 or more channels) provides the possibility to compare chemotactic activity of different cell populations or study preference between ligands.

Counting of cells: positive responder cells could be counted from the front of migrating cells, after staining or in native conditions in light microscope.

Two-chamber Techniques

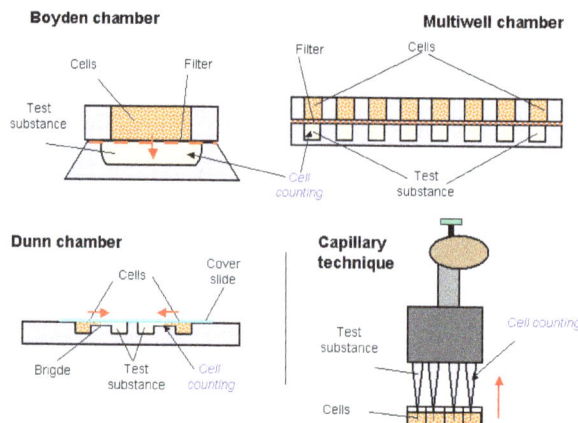

Chemotaxis chamber assays

Boyden Chamber

Chambers isolated by filters are proper tools for accurate determination of chemotactic behavior. The pioneer type of these chambers was constructed by Boyden. The motile cells are placed into the upper chamber, while fluid containing the test substance is filled into the lower one. The size of the motile cells to be investigated determines the pore size of the filter; it is essential to choose a diameter which allows active transmigration. For modelling *in vivo* conditions, several protocols prefer coverage of filter with molecules of extracellular matrix (collagen, elastin etc.) Efficiency of the measurements was increased by development of multiwell chambers (e.g. NeuroProbe), where 24, 96, 384 samples are evaluated in parallel. Advantage of this variant is that several parallels are assayed in identical conditions.

Bridge Chambers

In another setting the chambers are connected side by side horizontally (Zigmond chamber) or as concentric rings on a slide (Dunn chamber) Concentration gradient develops on a narrow connecting bridge between the chambers and the number of migrating cells is also counted on the surface of the bridge by light microscope. In some cases the bridge between the two chambers is filled with agar and cells have to "glide" in this semisolid layer.

Capillary Techniques

Some capillary techniques provide also a chamber like arrangement, however, there is no filter between the cells and the test substance. Quantitative results are gained by the multiwell type of this probe using 4-8-12-channel pipettes. Accuracy of the pipette and increased number of the parallel running samples is the great advantage of this test.

Counting of cells: positive responder cells are count from the lower chamber (long incubation time) or from the filter (short incubation time). For detection of cells general staining techniques (e.g. trypan blue) or special probes (e.g. mt-dehydrogenase detection with MTT assay) are used. Labelled (e.g. fluorochromes) cells are also used, in some assays cells get labelled during transmigration the filter.

Other Techniques

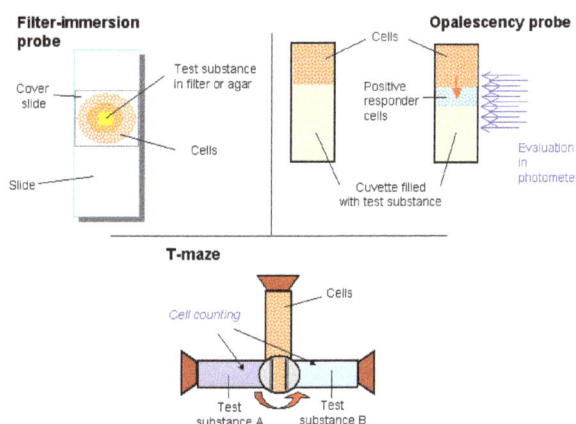

Other chemotaxis assay techniques

Besides the above-mentioned two, most commonly used family of techniques a wide range of protocols were developed to measure chemotactic activity. Some of them are only qualitative, like aggregation tests, where small pieces of agar or filters are placed onto a slide and accumulation of cells around is measured.

In another semiquantitative technique cells are overlaid the test substance and changes in opalescence of the originally cell-free compartment is recorded during the incubation time.

The third very frequently used, however, qualitative technique is the T-maze and its adaptations for microplates. In the original version a container drilled in a peg is filled with cells. Then the peg is twisted and the cells get contact with two other containers filled with different substances. The incubation is stopped with resetting the peg, the cell number is counted from the containers.

Coagulation

Coagulation (also known as clotting) is the process by which blood changes from a liquid to a gel, forming a blood clot. It potentially results in hemostasis, the cessation of blood loss from a damaged vessel, followed by repair. The mechanism of coagulation involves activation, adhesion, and aggregation of platelets along with deposition and maturation of fibrin. Disorders of coagulation are disease states which can result in bleeding (hemorrhage or bruising) or obstructive clotting (thrombosis).

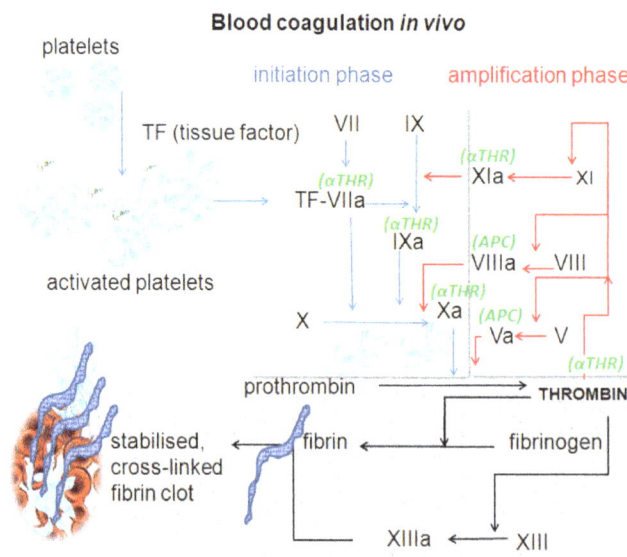

Blood coagulation pathways in vivo showing the central role played by thrombin

Coagulation is highly conserved throughout biology; in all mammals, coagulation involves both a cellular (platelet) and a protein (coagulation factor) component. The system in humans has been the most extensively researched and is the best understood.

Coagulation begins almost instantly after an injury to the blood vessel has damaged the endothelium lining the vessel. Leaking of blood through the endothelium initiates two processes: changes in platelets, and the exposure of subendothilial tissue factor to plasma Factor VII, which ultimately leads to fibrin formation. Platelets immediately form a plug at the site of injury; this is called *primary hemostasis*. *Secondary hemostasis* occurs simultaneously: Additional coagulation factors or clotting factors beyond Factor VII (listed below) respond in a complex cascade to form fibrin strands, which strengthen the platelet plug.

Physiology

Platelet Activation

When the endothelium is damaged, the normally isolated, underlying collagen is exposed to circulating platelets, which bind directly to collagen with collagen-specific glycoprotein Ia/IIa surface receptors. This adhesion is strengthened further by von Willebrand factor (vWF), which is released from the endothelium and from platelets; vWF forms additional links between the platelets' glycoprotein Ib/IX/V and the collagen fibrils. This localization of platelets to the extracellular matrix promotes collagen interaction with platelet glycoprotein VI. Binding of collagen to glycoprotein VI triggers a signaling cascade that results in activation of platelet integrins. Activated integrins mediate tight binding of platelets to the extracellular matrix. This process adheres platelets to the site of injury.

Activated platelets will release the contents of stored granules into the blood plasma. The granules include ADP, serotonin, platelet-activating factor (PAF), vWF, platelet factor 4, and thromboxane A_2 (TXA_2), which, in turn, activate additional platelets. The granules' contents activate a G_q-linked protein receptor cascade, resulting in increased calcium concentration in the platelets' cytosol. The calcium activates protein kinase C, which, in turn, activates phospholipase A_2 (PLA_2). PLA_2 then modifies the integrin membrane glycoprotein IIb/IIIa, increasing its affinity to bind fibrinogen. The activated platelets change shape from spherical to stellate, and the fibrinogen cross-links with glycoprotein IIb/IIIa aid in aggregation of adjacent platelets (completing primary hemostasis).

Coagulation Cascade

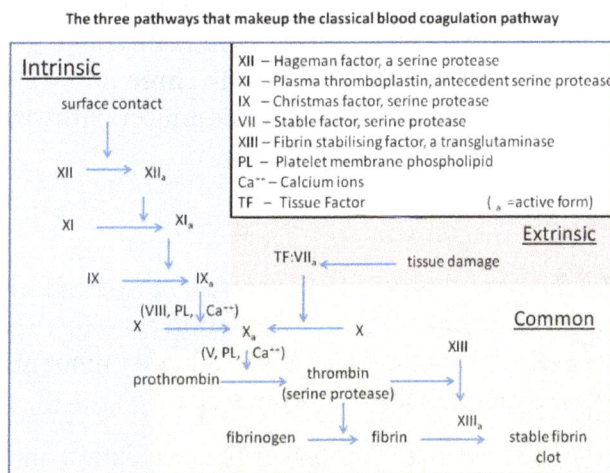

The classical blood coagulation pathway

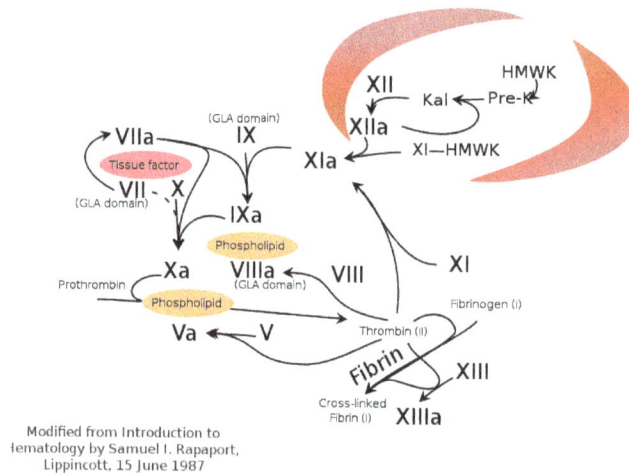

Modified from Introduction to
Hematology by Samuel I. Rapaport,
Lippincott, 15 June 1987

Modern coagulation pathway. Hand-drawn composite from similar drawings presented by Professor Dzung Le, MD, PhD, at UCSD Clinical Chemistry conferences on 14 and 21 October 2014. Original schema from Introduction to Hematology by Samuel I. Rapaport. 2nd edition;Lippencott:1987. Dr Le added the factor XI portion based on a paper from about year 2000. Dr Le's similar drawings presented the development of this cascade over 6 frames, like a comic.

The coagulation cascade of secondary hemostasis has two initial pathways which lead to *fibrin* formation. These are the *contact activation pathway* (also known as the intrinsic pathway), and the *tissue factor pathway* (also known as the extrinsic pathway) which both lead to the same fundamental reactions that produce fibrin. It was previously thought that the two pathways of coagulation cascade were of equal importance, but it is now known that the primary pathway for the initiation of blood coagulation is the *tissue factor* (extrinsic) pathway. The pathways are a series of reactions, in which a zymogen (inactive enzyme precursor) of a serine protease and its glycoprotein co-factor are activated to become active components that then catalyze the next reaction in the cascade, ultimately resulting in cross-linked fibrin. Coagulation factors are generally indicated by Roman numerals, with a lowercase *a* appended to indicate an active form.

The coagulation factors are generally serine proteases (enzymes), which act by cleaving downstream proteins. There are some exceptions. For example, FVIII and FV are glycoproteins, and Factor XIII is a transglutaminase. The coagulation factors circulate as inactive zymogens. The coagulation cascade is therefore classically divided into three pathways. The *tissue factor* and *contact activation* pathways both activate the "final common pathway" of factor X, thrombin and fibrin.

Tissue Factor Pathway (Extrinsic)

The main role of the tissue factor pathway is to generate a "thrombin burst", a process by which thrombin, the most important constituent of the coagulation cascade in terms of its feedback activation roles, is released very rapidly. FVIIa circulates in a higher amount than any other activated coagulation factor. The process includes the following steps:

1. Following damage to the blood vessel, FVII leaves the circulation and comes into contact with tissue factor (TF) expressed on tissue-factor-bearing cells (stromal fibroblasts and leukocytes), forming an activated complex (TF-FVIIa).

2. TF-FVIIa activates FIX and FX.

3. FVII is itself activated by thrombin, FXIa, FXII and FXa.

4. The activation of FX (to form FXa) by TF-FVIIa is almost immediately inhibited by tissue factor pathway inhibitor (TFPI).

5. FXa and its co-factor FVa form the prothrombinase complex, which activates prothrombin to thrombin.

6. Thrombin then activates other components of the coagulation cascade, including FV and FVIII (which forms a complex with FIX), and activates and releases FVIII from being bound to vWF.

7. FVIIIa is the co-factor of FIXa, and together they form the "tenase" complex, which activates FX; and so the cycle continues. ("Tenase" is a contraction of "ten" and the suffix "-ase" used for enzymes.)

Contact Activation Pathway (Intrinsic)

The contact activation pathway begins with formation of the primary complex on collagen by high-molecular-weight kininogen (HMWK), prekallikrein, and FXII (Hageman factor). Prekallikrein is converted to kallikrein and FXII becomes FXIIa. FXIIa converts FXI into FXIa. Factor XIa activates FIX, which with its co-factor FVIIIa form the tenase complex, which activates FX to FXa. The minor role that the contact activation pathway has in initiating clot formation can be illustrated by the fact that patients with severe deficiencies of FXII, HMWK, and prekallikrein do not have a bleeding disorder. Instead, contact activation system seems to be more involved in inflammation, and innate immunity. Despite this, interference with the pathway may confer protection against thrombosis without a significant bleeding risk.

Final Common Pathway

The division of coagulation in two pathways is mainly artificial, it originates from laboratory tests in which clotting times were measured after the clotting was initiated by glass (intrinsic pathway) or by thromboplastin (a mix of tissue factor and phospholipids). In fact thrombin is present from the very beginning, already when platelets are making the plug. *Thrombin* has a large array of functions, not only the conversion of fibrinogen to fibrin, the building block of a hemostatic plug. In addition, it is the most important platelet activator and on top of that it activates Factors VIII and V and their inhibitor protein C (in the presence of thrombomodulin), and it activates Factor XIII, which forms covalent bonds that crosslink the fibrin polymers that form from activated monomers.

Following activation by the contact factor or tissue factor pathways, the coagulation cascade is maintained in a prothrombotic state by the continued activation of FVIII and FIX to form the tenase complex, until it is down-regulated by the anticoagulant pathways.

Cofactors

Various substances are required for the proper functioning of the coagulation cascade:

Calcium and Phospholipid

Calcium and phospholipid (a platelet membrane constituent) are required for the tenase and pro-thrombinase complexes to function. Calcium mediates the binding of the complexes via the terminal gamma-carboxy residues on FXa and FIXa to the phospholipid surfaces expressed by platelets, as well as procoagulant microparticles or microvesicles shed from them. Calcium is also required at other points in the coagulation cascade.

Vitamin K

Vitamin K is an essential factor to a hepatic gamma-glutamyl carboxylase that adds a carboxyl group to glutamic acid residues on factors II, VII, IX and X, as well as Protein S, Protein C and Protein Z. In adding the gamma-carboxyl group to glutamate residues on the immature clotting factors Vitamin K is itself oxidized. Another enzyme, *Vitamin K epoxide reductase*, (VKORC) reduces vitamin K back to its active form. Vitamin K epoxide reductase is pharmacologically important as a target of anticoagulant drugs warfarin and related coumarins such as aceno-coumarol, phenprocoumon, and dicumarol. These drugs create a deficiency of reduced vitamin K by blocking VKORC, thereby inhibiting maturation of clotting factors. Vitamin K deficiency from other causes (e.g., in malabsorption) or impaired vitamin K metabolism in disease (e.g., in liver failure) lead to the formation of PIVKAs (proteins formed in vitamin K absence) which are partially or totally non-gamma carboxylated, affecting the coagulation factors' ability to bind to phospholipid.

Regulators

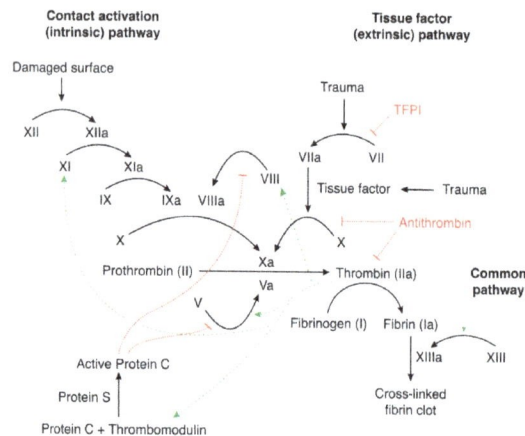

Coagulation with arrows for negative and positive feedback.

Five mechanisms keep platelet activation and the coagulation cascade in check. Abnormalities can lead to an increased tendency toward thrombosis:

Protein C

Protein C is a major physiological anticoagulant. It is a vitamin K-dependent serine protease enzyme that is activated by thrombin into activated protein C (APC). Protein C is activated in a sequence that starts with Protein C and thrombin binding to a cell surface protein thrombomodulin.

Thrombomodulin binds these proteins in such a way that it activates Protein C. The activated form, along with protein S and a phospholipid as cofactors, degrades FVa and FVIIIa. Quantitative or qualitative deficiency of either (protein C or protein S) may lead to thrombophilia (a tendency to develop thrombosis). Impaired action of Protein C (activated Protein C resistance), for example by having the "Leiden" variant of Factor V or high levels of FVIII also may lead to a thrombotic tendency.

Antithrombin

Antithrombin is a serine protease inhibitor (serpin) that degrades the serine proteases: thrombin, FIXa, FXa, FXIa, and FXIIa. It is constantly active, but its adhesion to these factors is increased by the presence of heparan sulfate (a glycosaminoglycan) or the administration of heparins (different heparinoids increase affinity to FXa, thrombin, or both). Quantitative or qualitative deficiency of antithrombin (inborn or acquired, e.g., in proteinuria) leads to thrombophilia.

Tissue Factor Pathway Inhibitor (TFPI)

Tissue factor pathway inhibitor (TFPI) limits the action of tissue factor (TF). It also inhibits excessive TF-mediated activation of FVII and FX.

Plasmin

Plasmin is generated by proteolytic cleavage of plasminogen, a plasma protein synthesized in the liver. This cleavage is catalyzed by tissue plasminogen activator (t-PA), which is synthesized and secreted by endothelium. Plasmin proteolytically cleaves fibrin into fibrin degradation products that inhibit excessive fibrin formation.

Prostacyclin

Prostacyclin (PGI_2) is released by endothelium and activates platelet G_s protein-linked receptors. This, in turn, activates adenylyl cyclase, which synthesizes cAMP. cAMP inhibits platelet activation by decreasing cytosolic levels of calcium and, by doing so, inhibits the release of granules that would lead to activation of additional platelets and the coagulation cascade.

Fibrinolysis

Eventually, blood clots are reorganised and resorbed by a process termed *fibrinolysis*. The main enzyme responsible for this process (plasmin) is regulated by various activators and inhibitors.

Role in Immune System

The coagulation system overlaps with the immune system. Coagulation can physically trap invading microbes in blood clots. Also, some products of the coagulation system can contribute to the innate immune system by their ability to increase vascular permeability and act as chemotactic agents for phagocytic cells. In addition, some of the products of the coagulation system are directly antimicrobial. For example, beta-lysine, an amino acid produced by platelets during coagulation, can cause lysis of many Gram-positive bacteria by acting as a cationic detergent. Many acute-phase proteins of inflammation are involved in the coagulation system. In addition, pathogenic bacteria may secrete agents that alter the coagulation system, e.g. coagulase and streptokinase.

Assessment

Numerous tests are used to assess the function of the coagulation system:

- Common: aPTT, PT (also used to determine INR), fibrinogen testing (often by the Clauss method), platelet count, platelet function testing (often by PFA-100), thrombodynamics test.

- Other: TCT, bleeding time, mixing test (whether an abnormality corrects if the patient's plasma is mixed with normal plasma), coagulation factor assays, antiphospholipid antibodies, D-dimer, genetic tests (e.g. factor V Leiden, prothrombin mutation G20210A), dilute Russell's viper venom time (dRVVT), miscellaneous platelet function tests, thromboelastography (TEG or Sonoclot), euglobulin lysis time (ELT).

The contact activation (intrinsic) pathway is initiated by activation of the "contact factors" of plasma, and can be measured by the activated partial thromboplastin time (aPTT) test.

The tissue factor (extrinsic) pathway is initiated by release of tissue factor (a specific cellular lipoprotein), and can be measured by the prothrombin time (PT) test. PT results are often reported as ratio (INR value) to monitor dosing of oral anticoagulants such as warfarin.

The quantitative and qualitative screening of fibrinogen is measured by the thrombin clotting time (TCT). Measurement of the exact amount of fibrinogen present in the blood is generally done using the Clauss method for fibrinogen testing. Many analysers are capable of measuring a "derived fibrinogen" level from the graph of the Prothrombin time clot.

If a coagulation factor is part of the contact activation or tissue factor pathway, a deficiency of that factor will affect only one of the tests: Thus hemophilia A, a deficiency of factor VIII, which is part of the contact activation pathway, results in an abnormally prolonged aPTT test but a normal PT test. The exceptions are prothrombin, fibrinogen, and some variants of FX that can be detected only by either aPTT or PT. If an abnormal PT or aPTT is present, additional testing will occur to determine which (if any) factor is present as aberrant concentrations.

Deficiencies of fibrinogen (quantitative or qualitative) will affect all screening tests.

Role in Disease

Coagulation defects may cause hemorrhage or thrombosis, and occasionally both, depending on the nature of the defect.

Platelet Disorders

Platelet conditions may be congenital or acquired. Some inborn platelet pathologies are Glanzmann's thrombasthenia, Bernard-Soulier syndrome (abnormal glycoprotein Ib-IX-V complex), gray platelet syndrome (deficient alpha granules), and delta storage pool deficiency (deficient dense granules). Most are rare conditions. Most inborn platelet pathologies predispose to hemorrhage. Von Willebrand disease is due to deficiency or abnormal function of von Willebrand factor, and leads to a similar bleeding pattern; its milder forms are relatively common.

Decreased platelet numbers may be due to various causes, including insufficient production (e.g., in myelodysplastic syndrome or other bone marrow disorders), destruction by the immune system (immune thrombocytopenic purpura/ITP), and consumption due to various causes (thrombotic thrombocytopenic purpura/TTP, hemolytic-uremic syndrome/HUS, paroxysmal nocturnal hemoglobinuria/PNH, disseminated intravascular coagulation/DIC, heparin-induced thrombocytopenia/HIT). Most consumptive conditions lead to platelet activation, and some are associated with thrombosis.

Disease and Clinical Significance of Thrombosis

The best-known coagulation factor disorders are the hemophilias. The three main forms are hemophilia A (factor VIII deficiency), hemophilia B (factor IX deficiency or "Christmas disease") and hemophilia C (factor XI deficiency, mild bleeding tendency). Hemophilia A and B are X-linked recessive disorders, whereas Hemophilia C is a much more rare autosomal recessive disorder most commonly seen in Ashkenazi Jews.

Von Willebrand disease (which behaves more like a platelet disorder except in severe cases), is the most common hereditary bleeding disorder and is characterized as being inherited autosomal recessive or dominant. In this disease, there is a defect in von Willebrand factor (vWF), which mediates the binding of glycoprotein Ib (GPIb) to collagen. This binding helps mediate the activation of platelets and formation of primary hemostasis.

Bernard-Soulier syndrome is a defect or deficiency in GPIb. GPIb, the receptor for vWF, can be defective and lead to lack of primary clot formation (primary hemostasis) and increased bleeding tendency. This is an autosomal recessive inherited disorder.

Thrombasthenia of Glanzmann and Naegeli (Glanzmann thrombasthenia) is extremely rare. It is characterized by a defect in GPIIb/IIIa fibrinogen receptor complex. When GPIIb/IIIa receptor is dysfunctional, fibrinogen cannot cross-link platelets, which inhibits primary hemostasis. This is an autosomal recessive inherited disorder.

In liver failure (acute and chronic forms), there is insufficient production of coagulation factors by the liver; this may increase bleeding risk.

Deficiency of Vitamin K may also contribute to bleeding disorders because clotting factor maturation depends on Vitamin K.

Thrombosis is the pathological development of blood clots. These clots may break free and become mobile, forming an embolus or grow to such a size that occludes the vessel in which it developed. An embolism is said to occur when the thrombus (blood clot) becomes a mobile embolus and migrates to another part of the body, interfering with blood circulation and hence impairing organ function downstream of the occlusion. This causes ischemia and often leads to ischemic necrosis of tissue. Most cases of venous thrombosis are due to acquired states (older age, surgery, cancer, immobility) or inherited thrombophilias (e.g., antiphospholipid syndrome, factor V Leiden, and various other genetic deficiencies or variants).

Mutations in factor XII have been associated with an asymptomatic prolongation in the clotting time and possibly a tendency toward thrombophlebitis. Other mutations have been linked with a rare form of hereditary angioedema (type III) essentialism.

Pharmacology

Procoagulants

The use of adsorbent chemicals, such as zeolites, and other hemostatic agents are also used for sealing severe injuries quickly (such as in traumatic bleeding secondary to gunshot wounds). Thrombin and fibrin glue are used surgically to treat bleeding and to thrombose aneurysms.

Desmopressin is used to improve platelet function by activating arginine vasopressin receptor 1A.

Coagulation factor concentrates are used to treat hemophilia, to reverse the effects of anticoagulants, and to treat bleeding in patients with impaired coagulation factor synthesis or increased consumption. Prothrombin complex concentrate, cryoprecipitate and fresh frozen plasma are commonly used coagulation factor products. Recombinant activated human factor VII is increasingly popular in the treatment of major bleeding.

Tranexamic acid and aminocaproic acid inhibit fibrinolysis, and lead to a *de facto* reduced bleeding rate. Before its withdrawal, aprotinin was used in some forms of major surgery to decrease bleeding risk and need for blood products.

Anticoagulants

Anticoagulants and anti-platelet agents are amongst the most commonly used medications. Anti-platelet agents include aspirin, dipyridamole, ticlopidine, clopidogrel, ticagrelor and prasugrel; the parenteral glycoprotein IIb/IIIa inhibitors are used during angioplasty. Of the anticoagulants, warfarin (and related coumarins) and heparin are the most commonly used. Warfarin affects the vitamin K-dependent clotting factors (II, VII, IX,X) and protein C and protein S, whereas heparin and related compounds increase the action of antithrombin on thrombin and factor Xa. A newer class of drugs, the direct thrombin inhibitors, is under development; some members are already in clinical use (such as lepirudin). Also under development are other small molecular compounds that interfere directly with the enzymatic action of particular coagulation factors (e.g., rivaroxaban, dabigatran, apixaban).

Coagulation Factors

Coagulation factors and related substances		
Number and/or name	**Function**	**Associated genetic disorders**
I (fibrinogen)	Forms clot (fibrin)	Congenital afibrinogenemia, Familial renal amyloidosis
II (prothrombin)	Its active form (IIa) activates I, V, X, VII, VIII, XI, XIII, protein C, platelets	Prothrombin G20210A, Thrombophilia

III (tissue factor or tissue thromboplastin)	Co-factor of VIIa (formerly known as factor III)	
IV Calcium	Required for coagulation factors to bind to phospholipid (formerly known as factor IV)	
V (proaccelerin, labile factor)	Co-factor of X with which it forms the prothrombinase complex	Activated protein C resistance
VI	*Unassigned* – old name of Factor Va	
VII (stable factor, proconvertin)	Activates IX, X	congenital factor VII deficiency
VIII (Antihemophilic factor A)	Co-factor of IX with which it forms the tenase complex	Haemophilia A
IX (Antihemophilic factor B or Christmas factor)	Activates X: forms tenase complex with factor VIII	Haemophilia B
X (Stuart-Prower factor)	Activates II: forms prothrombinase complex with factor V	Congenital Factor X deficiency
XI (plasma thromboplastin antecedent)	Activates IX	Haemophilia C
XII (Hageman factor)	Activates factor XI, VII and prekallikrein	Hereditary angioedema type III
XIII (fibrin-stabilizing factor)	Crosslinks fibrin	Congenital Factor XIIIa/b deficiency
von Willebrand factor	Binds to VIII, mediates platelet adhesion	von Willebrand disease
prekallikrein (Fletcher factor)	Activates XII and prekallikrein; cleaves HMWK	Prekallikrein/Fletcher Factor deficiency
high-molecular-weight kininogen (HMWK) (Fitzgerald factor)	Supports reciprocal activation of XII, XI, and prekallikrein	Kininogen deficiency
fibronectin	Mediates cell adhesion	Glomerulopathy with fibronectin deposits
antithrombin III	Inhibits IIa, Xa, and other proteases	Antithrombin III deficiency
heparin cofactor II	Inhibits IIa, cofactor for heparin and dermatan sulfate ("minor antithrombin")	Heparin cofactor II deficiency
protein C	Inactivates Va and VIIIa	Protein C deficiency
protein S	Cofactor for activated protein C (APC, inactive when bound to C4b-binding protein)	Protein S deficiency
protein Z	Mediates thrombin adhesion to phospholipids and stimulates degradation of factor X by ZPI	Protein Z deficiency
Protein Z-related protease inhibitor (ZPI)	Degrades factors X (in presence of protein Z) and XI (independently)	
plasminogen	Converts to plasmin, lyses fibrin and other proteins	Plasminogen deficiency, type I (ligneous conjunctivitis)
alpha 2-antiplasmin	Inhibits plasmin	Antiplasmin deficiency
tissue plasminogen activator (tPA)	Activates plasminogen	Familial hyperfibrinolysis and thrombophilia
urokinase	Activates plasminogen	Quebec platelet disorder
plasminogen activator inhibitor-1 (PAI1)	Inactivates tPA & urokinase (endothelial PAI)	Plasminogen activator inhibitor-1 deficiency
plasminogen activator inhibitor-2 (PAI2)	Inactivates tPA & urokinase (placental PAI)	
cancer procoagulant	Pathological factor X activator linked to thrombosis in cancer	

History

Initial Discoveries

Theories on the coagulation of blood have existed since antiquity. Physiologist Johannes Müller (1801–1858) described fibrin, the substance of a thrombus. Its soluble precursor, fibrinogen, was thus named by Rudolf Virchow (1821–1902), and isolated chemically by Prosper Sylvain Denis (1799–1863). Alexander Schmidt suggested that the conversion from fibrinogen to fibrin is the result of an enzymatic process, and labeled the hypothetical enzyme "thrombin" and its precursor "prothrombin". Arthus discovered in 1890 that calcium was essential in coagulation. Platelets were identified in 1865, and their function was elucidated by Giulio Bizzozero in 1882.

The theory that thrombin is generated by the presence of tissue factor was consolidated by Paul Morawitz in 1905. At this stage, it was known that *thrombokinase/thromboplastin* (factor III) is released by damaged tissues, reacting with *prothrombin* (II), which, together with calcium (IV), forms *thrombin*, which converts fibrinogen into *fibrin* (I).

Coagulation Factors

The remainder of the biochemical factors in the process of coagulation were largely discovered in the 20th century.

A first clue as to the actual complexity of the system of coagulation was the discovery of *proaccelerin* (initially and later called Factor V) by Paul Owren (1905–1990) in 1947. He also postulated its function to be the generation of accelerin (Factor VI), which later turned out to be the activated form of V (or Va); hence, VI is not now in active use.

Factor VII (also known as *serum prothrombin conversion accelerator* or *proconvertin*, precipitated by barium sulfate) was discovered in a young female patient in 1949 and 1951 by different groups.

Factor VIII turned out to be deficient in the clinically recognised but etiologically elusive hemophilia A; it was identified in the 1950s and is alternatively called *antihemophilic globulin* due to its capability to correct hemophilia A.

Factor IX was discovered in 1952 in a young patient with hemophilia B named Stephen Christmas (1947–1993). His deficiency was described by Dr. Rosemary Biggs and Professor R.G. MacFarlane in Oxford, UK. The factor is, hence, called Christmas Factor. Christmas lived in Canada, and campaigned for blood transfusion safety until succumbing to transfusion-related AIDS at age 46. An alternative name for the factor is *plasma thromboplastin component*, given by an independent group in California.

Hageman factor, now known as factor XII, was identified in 1955 in an asymptomatic patient with a prolonged bleeding time named of John Hageman. Factor X, or Stuart-Prower factor, followed, in 1956. This protein was identified in a Ms. Audrey Prower of London, who had a lifelong bleeding tendency. In 1957, an American group identified the same factor in a Mr. Rufus Stuart. Factors XI and XIII were identified in 1953 and 1961, respectively.

The view that the coagulation process is a "cascade" or "waterfall" was enunciated almost simultaneously by MacFarlane in the UK and by Davie and Ratnoff in the USA, respectively.

Nomenclature

The usage of Roman numerals rather than eponyms or systematic names was agreed upon during annual conferences (starting in 1955) of hemostasis experts. In 1962, consensus was achieved on the numbering of factors I-XII. This committee evolved into the present-day International Committee on Thrombosis and Hemostasis (ICTH). Assignment of numerals ceased in 1963 after the naming of Factor XIII. The names Fletcher Factor and Fitzgerald Factor were given to further coagulation-related proteins, namely prekallikrein and high-molecular-weight kininogen, respectively.

Factors III and VI are unassigned, as thromboplastin was never identified, and actually turned out to consist of ten further factors, and accelerin was found to be activated Factor V.

Other Species

All mammals have an extremely closely related blood coagulation process, using a combined cellular and serine protease process. In fact, it is possible for any mammalian coagulation factor to "cleave" its equivalent target in any other mammal. The only non-mammalian animal known to use serine proteases for blood coagulation is the horseshoe crab.

Stealth Characteristics in Nanoparticles

It is evident from the various immune recognition processes that protein adsorption is one of the prime factors that results in activation of the immune response. Hence, any strategy to avoid immune response should be directed in interfering with the protein adsorption process.

Dysopsonins

Just as the adsorption of the opsonin proteins promote immune system activation, there is another category of proteins found in the blood known as dysopsonins that have the ability to promote the desorption of the opsonin proteins from the surface. If the dysopsonins adsorb on to the surface, further adsorption of the opsonins is inhibited. A few serum DNA binding proteins (SDBP) have been implicated to possess dysopsonin activity. However, the dysopsonins present in the blood are yet to positively identified thus limiting the scope of application of this strategy. Moreover, it remains to be confirmed whether the introduction of a dysopsonin protein from a particular species in another species can stimulate antibody production against it.

Role of Ions in Immune Recognition

In several trials involving nanocarriers of the same size, shape and chemical nature, the magnitude of opsonization on the nanocarriers was measured in the presence of citrate ions and in the absence of citrate ions. It was found that the levels of protein adsorption sharply decreased in the presence of citrate ions. What could be the reason? Citrate is a good chelator of ions, especially the divalent ions. It is therefore evident that when citrate was present in the medium, all divalent ions such as Ca^{2+} and Mg^{2+} were chelated. This implies that the opsonization, complement activation and coagulation process is dependent on the presence of divalent ions in the medium. Though this experiment provides excellent insight into the importance of divalent ions in the immune system activation, the feasibility of utilizing this property for imparting stealth characteristics is poor. This is because chelating divalent ions in vivo will lead to the disruption of the fragile ion balance that can cause undesirable effects on other metabolic functions of the system.

Role of Hydroxyl Functionalities in Immune Recognition

Zymosan, a polysaccharide from yeast, when introduced into the biological system generates a strong immune response through complement activation. Similarly, membranes made from regenerated cellulose have also induced high levels of complement activation. Sephadex or cross-linked dextran also has been found to activate the complement. What is common between these three examples apart from the fact that they all belong to the polysaccharide family? All of them contain a large number of hydroxyl groups (-OH). This suggests that presence of abundant surface –OH groups might serve to elicit a high degree of immune response through complement activation. Indeed, it was observed that hydroxymethyl polystyrene containing over 90% hydroxyl groups strongly activated the complement. On the other hand substitution of the hydroxyl groups in sephadex with diethylaminoethyl (DEAE) groups significantly reduced the magnitude of complement activation. This is on expected lines since the alternative pathway for activation of the complement involves lysis of the thioester bond in C3b- like-C3 and formation of an ester link with –OH groups on the surface of the foreign body.

Intriguingly, polystyrene, which does not contain any hydroxyl groups also activates the complement. This indicates that the presence of hydroxyl groups on the surface may not be the main reason for complement activation; rather it may serve to enhance the immune response once activated! This highlights the fact that the nature and extent of protein adsorption on the surface of the nanocarrier may have a major role in determining the magnitude of immune response. Therefore, the right strategy would be to inhibit the protein adsorption on the surface.

Inhibition of Opsonization by Steric Repulsion

The non-ionic surfactant Pluronics® (also known as Poloxamer) when coated on nanoparticles was found to reduce the protein adsorption and hence exhibit a marked decrease in the complement activation. Pluronics® is a tri-block copolymer of poly(ethylene oxide)-co-poly(propylene oxide)-co-poly(ethylene oxide). Figure shows the structure of this polymer:

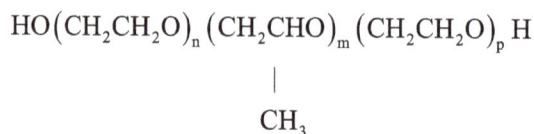

$$HO(CH_2CH_2O)_n (CH_2CHO)_m (CH_2CH_2O)_p H$$
$$|$$
$$CH_3$$

The high hydrophilicity and chain mobility of the poly(ethylene oxide) segments could be instrumental in conferring protein repulsion properties. Poly(ethylene oxide) has hydrophilic character that tends to attract water molecules around itself. Also, the poly(ethylene oxide) segments possess the ability to move about due to the small groups in their chain. A constantly moving segment with a layer of water molecules presents a surface on which protein adsorption becomes difficult.

When nanocarriers were coated with poly(ethylene glycol) that have the same structure as poly(ethylene oxide) but only differ in the molecular weight, protein adsorption was significantly reduced indicating the hydrophilicity and fast chain dynamics are both essential to provide protection against opsonization processes. Surfaces that possess only hydrophilic nature but lack chain mobility (as in the case of sephadex), exhibit very high immune response. The discovery of the protein-repelling property of poly(ethylene glycol) (PEG) has been widely exploited in developing drug delivery systems that exhibit considerably less degree of immune activation owing to the retardation in opsonization. The process of introducing poly(ethylene glycol) to a nanocarrier is termed as 'PEGylation'.

PEGylation

PEGylation (often styled pegylation) is the process of both covalent and non-covalent attachment or amalgamation of polyethylene glycol (PEG, in pharmacy called macrogol) polymer chains to molecules and macrostructures, such as a drug, therapeutic protein or vesicle, which is then described as PEGylated (pegylated). PEGylation is routinely achieved by incubation of a reactive derivative of PEG with the target molecule. The covalent attachment of PEG to a drug or therapeutic protein can "mask" the agent from the host's immune system (reduced immunogenicity and antigenicity), and increase the hydrodynamic size (size in solution) of the agent which prolongs its circulatory time by reducing renal clearance. PEGylation can also provide water solubility to hydrophobic drugs and proteins.

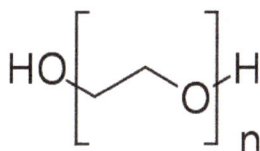

$$HO \left[\diagup\diagdown_O \right]_n H$$

Polyethylene glycol (in pharmacy called macrogol)

History

Around 1970, Frank F. Davis, a professor of biochemistry at Rutgers University, became interested in developing a process to render usable bioactive proteins of potential medical value. After considerable study, he concluded that the attachment of an inert and hydrophilic polymer might extend blood life and control immunogenicity of the proteins. Polyethylene glycol was chosen as the polymer. A team of Davis, Theodorus Van Es and Nicholas C. Palczuk conducted animal studies and found that PEG attachment greatly extended blood life and controlled immunogenicity of the proteins. A patent application was filed in 1973 and patent issued in 1979. The inventors and Abraham Abuchowski conducted extensive additional PEGylation studies on various enzymes. In 1981 Davis and Abuchowski founded Enzon, Inc., which brought three PEGylated drugs to market. Abuchowski later founded and is CEO of Prolong Pharmaceuticals.

Overview

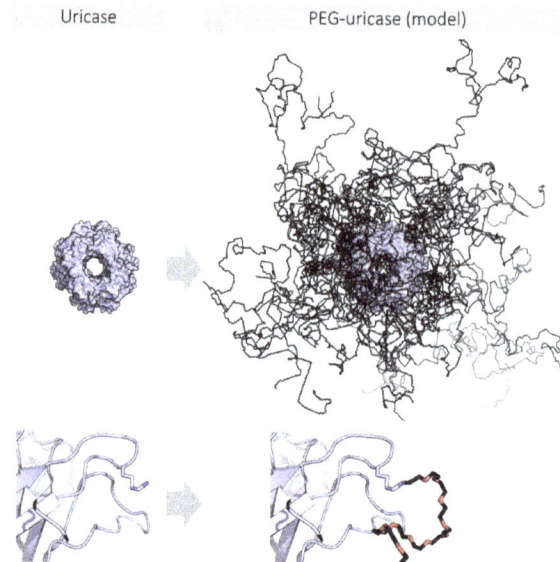

A comparison of uricase and PEG-uricase. PEG-uricase includes 40 polymers of 10kDa PEG. PEGylation improves its solubility at physiological pH, increases serum half-life and reduces immunogenicity without compromising activity. Upper images show the whole tetramer, lower images show one of the lysines that is PEGylated. (uricase from PDB: 1uox and PEG-uricase model from reference; only 36 PEG polymers included)

PEGylation is the process of attaching the strands of the polymer PEG to molecules, most typically peptides, proteins, and antibody fragments, that can improve the safety and efficiency of many therapeutics. It produces alterations in the physiochemical properties including changes in conformation, electrostatic binding, hydrophobicity etc. These physical and chemical changes increase systemic retention of the therapeutic agent. Also, it can influence the binding affinity of the therapeutic moiety to the cell receptors and can alter the absorption and distribution patterns.

PEGylation, by increasing the molecular weight of a molecule, can impart several significant pharmacological advantages over the unmodified form, such as:

- Improved drug solubility

- Reduced dosage frequency, without diminished efficacy with potentially reduced toxicity

- Extended circulating life

- Increased drug stability

- Enhanced protection from proteolytic degradation

PEGylated drugs also have the following commercial advantages:

- Opportunities for new delivery formats and dosing regimens

- Extended patent life of previously approved drugs

PEG is a particularly attractive polymer for conjugation. The specific characteristics of PEG moieties relevant to pharmaceutical applications are:

- Water solubility

- High mobility in solution

- Lack of toxicity and low immunogenicity

- Ready clearance from the body

- Altered distribution in the body

PEGylated Pharmaceuticals on the Market

The clinical value of PEGylation is now well established. ADAGEN (pegademase bovine) manufactured by Enzon Pharmaceuticals, Inc., US was the first PEGylated protein approved by the U.S. Food and Drug Administration (FDA) in March 1990, to enter the market. It is used to treat X-linked severe combined immunogenicity syndrome, as an alternative to bone marrow transplantation and enzyme replacement by gene therapy. Since the introduction of ADAGEN, a large number of PEGylated protein and peptide pharmaceuticals have followed and many others are under clinical trial or under development stages. Sales of the two most successful products, Pegasys and Neulasta, exceeded $5 billion in 2011. All commercially available PEGylated pharmaceuticals contain methoxypoly(ethylene glycol) or mPEG. PEGylated pharmaceuticals currently on the market (in reverse chronology by FDA approval year) include:

- Adynovate — PEGylated Antihemophilic Factor VIII for the treatment of patients with Hemophilia A. (Baxalta, 2015)

- Plegridy — PEGylated Interferon Beta-1a for the treatment of patients with relapsing forms of multiple sclerosis. (Biogen, 2014)

- Naloxegol (Movantik) — PEGylated naloxol for the treatment of opioid-induced constipation in adults patients with chronic non-cancer pain (un-pegylated methadone can cause adverse gastrointestinal reactions). (AstraZeneca, 2014)

- Peginesatide (Omontys) — once-monthly medication to treat anemia associated with chronic kidney disease in adult patients on dialysis (Affymax/Takeda Pharmaceuticals, 2012)

- Pegloticase (Krystexxa) — PEGylated uricase for the treatment of gout (Savient, 2010)

- Certolizumab pegol (Cimzia) — monoclonal antibody for treatment of moderate to severe rheumatoid arthritis and Crohn's disease, an inflammatory gastrointestinal disorder (Nektar/UCB Pharma, 2008)

- Methoxy polyethylene glycol-epoetin beta (Mircera) — PEGylated form of erythropoetin to combat anemia associated with chronic kidney disease (Roche, 2007)

- Pegaptanib (Macugen) — used to treat neovascular age-related macular degeneration (Pfizer, 2004)

- Pegfilgrastim (Neulasta) — PEGylated recombinant methionyl human granulocyte colony-stimulating factor for severe cancer chemotherapy-induced neutropenia (Amgen, 2002)

- Pegvisomant (Somavert) — PEG-human growth hormone mutein antagonist for treatment of Acromegaly (Pfizer, 2002)

- Peginterferon alfa-2a (Pegasys) — PEGylated interferon alpha for use in the treatment of chronic hepatitis C and hepatitis B (Hoffmann-La Roche, 2001)

- Doxorubicin HCl liposome (Doxil/Caelyx) — PEGylated liposome containing doxorubicin for the treatment of cancer (Ortho Biotech/Schering-Plough, 2001)

- Peginterferon alfa-2b (PegIntron) — PEGylated interferon alpha for use in the treatment of chronic hepatitis C and hepatitis B (Schering-Plough/Enzon, 2000)

- Pegaspargase (Oncaspar) — PEGylated L-asparaginase for the treatment of acute lympho-blastic leukemia in patients who are hypersensitive to the native unmodified form of L-asparaginase (Enzon, 1994). This drug was recently approved for front line use.

- Pegademase bovine (Adagen) — PEG-adenosine deaminase for the treatment of severe combined immunodeficiency disease (SCID) (Enzon, 1990)

PEGylation Process

The first step of the PEGylation is the suitable functionalization of the PEG polymer at one or both terminals. PEGs that are activated at each terminus with the same reactive moiety are known as "homobifunctional", whereas if the functional groups present are different, then the PEG derivative is referred as "heterobifunctional" or "heterofunctional." The chemically active or activated derivatives of the PEG polymer are prepared to attach the PEG to the desired molecule.

The overall PEGylation processes used to date for protein conjugation can be broadly classified into two types, namely a solution phase batch process and an on-column fed-batch process. The simple and commonly adopted batch process involves the mixing of reagents together in a suitable buffer solution, preferably at a temperature between 4 and 6 °C, followed by the separation and purification of the desired product using a suitable technique based on its physicochemical properties, including size exclusion chromatography (SEC), ion exchange chromatography (IEX), hydrophobic interaction chromatography (HIC) and membranes or aqueous two phase systems.

The choice of the suitable functional group for the PEG derivative is based on the type of available reactive group on the molecule that will be coupled to the PEG. For proteins, typical reactive amino acids include lysine, cysteine, histidine, arginine, aspartic acid, glutamic acid, serine, threonine, tyrosine. The N-terminal amino group and the C-terminal carboxylic acid can also be used as a site specific site by conjugation with aldehyde functional polymers.

The techniques used to form first generation PEG derivatives are generally reacting the PEG polymer with a group that is reactive with hydroxyl groups, typically anhydrides, acid chlorides, chloroformates and carbonates. In the second generation PEGylation chemistry more efficient functional groups such as aldehyde, esters, amides etc. made available for conjugation.

As applications of PEGylation have become more and more advanced and sophisticated, there has been an increase in need for heterobifunctional PEGs for conjugation. These heterobifunctional PEGs are very useful in linking two entities, where a hydrophilic, flexible and biocompatible spacer is needed. Preferred end groups for heterobifunctional PEGs are maleimide, vinyl sulfones, pyridyl disulfide, amine, carboxylic acids and NHS esters.

Third generation pegylation agents, where the shape of the polymer has been branched, Y shaped or comb shaped are available which show reduced viscosity and lack of organ accumulation.

Limitations of PEGylation

Unpredictability in clearance times for PEGylated compounds may lead to the accumulation of large molecular weight compounds in the liver leading to inclusion bodies with no known toxicologic consequences. Furthermore, alteration in the chain length may lead to unexpected clearance times *in vivo*.

Future Perspectives

Four decades of development in PEGylation technology have proven its pharmacological advantages and acceptability. As a multibillion-dollar annual business with growing interest from both emerging biotechnology and established multinational pharmaceutical companies, there is great scientific and commercial interest in improving present methodologies and in introducing innovative process variations.

Polyethylene Glycol

Polyethylene glycol (PEG) is a polyether compound with many applications from industrial manufacturing to medicine. PEG is also known as polyethylene oxide (PEO) or polyoxyethylene (POE), depending on its molecular weight. The structure of PEG is commonly expressed as $H-(O-CH_2-CH_2)_n-OH$.

Available Forms and Nomenclature

PEG, *PEO*, and *POE* refer to an oligomer or polymer of ethylene oxide. The three names are chemically synonymous, but historically *PEG* is preferred in the biomedical field, whereas *PEO* is more prevalent in the field of polymer chemistry. Because different applications require different polymer chain lengths, *PEG* has tended to refer to oligomers and polymers with a molecular mass below 20,000 g/mol, *PEO* to polymers with a molecular mass above 20,000 g/mol, and *POE* to a polymer of any molecular mass. PEG and PEO are liquids or low-melting solids, depending on their molecular weights. PEGs are prepared by polymerization of ethylene oxide and are commercially available over a wide range of molecular weights from 300 g/mol to 10,000,000 g/mol. While PEG and PEO with different molecular weights find use in different applications, and have different physical properties (e.g. viscosity) due to chain length effects, their chemical properties are nearly identical. Different forms of PEG are also available, depending on the initiator used for the polymerization process – the most common initiator is a monofunctional methyl ether PEG, or methoxypoly(ethylene glycol), abbreviated mPEG. Lower-molecular-weight PEGs are also available as purer oligomers, referred to as monodisperse, uniform, or discrete. Very high purity PEG has recently been shown to be crystalline, allowing determination of a crystal structure by x-ray diffraction. Since purification and separation of pure oligomers is difficult, the price for this type of quality is often 10–1000 fold that of polydisperse PEG.

PEGs are also available with different geometries.

- *Branched* PEGs have three to ten PEG chains emanating from a central core group.

- *Star* PEGs have 10 to 100 PEG chains emanating from a central core group.

- *Comb* PEGs have multiple PEG chains normally grafted onto a polymer backbone.

The numbers that are often included in the names of PEGs indicate their average molecular weights (e.g. a PEG with n = 9 would have an average molecular weight of approximately 400 daltons, and would be labeled PEG 400.) Most PEGs include molecules with a distribution of molecular weights (i.e. they are polydisperse). The size distribution can be characterized statistically by its weight average molecular weight (Mw) and its number average molecular weight (Mn), the ratio of which is called the polydispersity index (Mw/Mn). MW and Mn can be measured by mass spectrometry.

PEGylation is the act of covalently coupling a PEG structure to another larger molecule, for example, a therapeutic protein, which is then referred to as a *PEGylated* protein. PEGylated interferon alfa-2a or –2b are commonly used injectable treatments for hepatitis C infection.

PEG is soluble in water, methanol, ethanol, acetonitrile, benzene, and dichloromethane, and is insoluble in diethyl ether and hexane. It is coupled to hydrophobic molecules to produce non-ionic surfactants.

PEGs potentially contain toxic impurities, such as ethylene oxide and 1,4-dioxane. Ethylene Glycol and its ethers are nephrotoxic if applied to damaged skin.

Polyethylene oxide (PEO, M_w 4 kDa) nanometric crystallites (4 nm)

Polyethylene glycol (PEG) and related polymers (PEG phospholipid constructs) are often sonicated when used in biomedical applications. However, as reported by Murali et al., PEG is very sensitive to sonolytic degradation and PEG degradation products can be toxic to mammalian cells. It is, thus, imperative to assess potential PEG degradation to ensure that the final material does not contain undocumented contaminants that can introduce artifacts into experimental results.

PEGs and methoxypolyethylene glycols are manufactured by Dow Chemical under the tradename *Carbowax* for industrial use, and *Carbowax Sentry* for food and pharmaceutical use. They vary in consistency from liquid to solid, depending on the molecular weight, as indicated by a number following the name. They are used commercially in numerous applications, including as surfactants, in foods, in cosmetics, in pharmaceutics, in biomedicine, as dispersing agents, as solvents, in ointments, in suppository bases, as tablet excipients, and as laxatives. Some specific groups are lauromacrogols, nonoxynols, octoxynols, and poloxamers.

Macrogol, used as a laxative, is a form of polyethylene glycol. The name may be followed by a number which represents the average molecular weight (e.g. macrogol 3350, macrogol 4000 or macrogol 6000).

Production

Polyethylene glycol 400, pharmaceutical quality

Polyethylene glycol 4000, pharmaceutical quality

Polyethylene glycol is produced by the interaction of ethylene oxide with water, ethylene glycol, or ethylene glycol oligomers. The reaction is catalyzed by acidic or basic catalysts. Ethylene glycol and its oligomers are preferable as a starting material instead of water, because they allow the creation of polymers with a low polydispersity (narrow molecular weight distribution). Polymer chain length depends on the ratio of reactants.

$$HOCH_2CH_2OH + n(CH_2CH_2O) \rightarrow HO(CH_2CH_2O)_{n+1}H$$

Depending on the catalyst type, the mechanism of polymerization can be cationic or anionic. The anionic mechanism is preferable because it allows one to obtain PEG with a low polydispersity.

Polymerization of ethylene oxide is an exothermic process. Overheating or contaminating ethylene oxide with catalysts such as alkalis or metal oxides can lead to runaway polymerization, which can end in an explosion after a few hours.

Polyethylene oxide, or high-molecular weight polyethylene glycol, is synthesized by suspension polymerization. It is necessary to hold the growing polymer chain in solution in the course of the polycondensation process. The reaction is catalyzed by magnesium-, aluminium-, or calcium-organoelement compounds. To prevent coagulation of polymer chains from solution, chelating additives such as dimethylglyoxime are used.

Alkaline catalysts such as sodium hydroxide (NaOH), potassium hydroxide (KOH), or sodium carbonate (Na_2CO_3) are used to prepare low-molecular-weight polyethylene glycol.

Uses

Medical uses

- PEG is the basis of a number of laxatives. Whole bowel irrigation with polyethylene glycol and added electrolytes is used for bowel preparation before surgery or colonoscopy.

- PEG is also used as an excipient in many pharmaceutical products.

- When attached to various protein medications, polyethylene glycol allows a slowed clearance of the carried protein from the blood.

The remains of the 16th century carrack *Mary Rose* undergoing conservation treatment with PEG in the 1980s

Chemical Uses

- Polyethylene glycol has a low toxicity and is used in a variety of products. The polymer is used as a lubricating coating for various surfaces in aqueous and non-aqueous environments.

- Since PEG is a flexible, water-soluble polymer, it can be used to create very high osmotic pressures (on the order of tens of atmospheres). It also is unlikely to have specific interactions with biological chemicals. These properties make PEG one of the most useful molecules for applying osmotic pressure in biochemistry and biomembranes experiments, in particular when using the osmotic stress technique.

- Polyethylene glycol is also commonly used as a polar stationary phase for gas chromatography, as well as a heat transfer fluid in electronic testers.

- PEG has also been used to preserve objects that have been salvaged from underwater, as was the case with the warship *Vasa* in Stockholm, the Mary Rose in England, the Ma'agan Michael Ship in Israel, and artifacts from the Steamboat Arabia in Kansas City, Missouri. It replaces water in wooden objects, making the wood dimensionally stable and preventing warping or shrinking of the wood when it dries. In addition, PEG is used when working with green wood as a stabilizer, and to prevent shrinkage.

- PEG has been used to preserve the painted colors on Terra-Cotta Warriors unearthed at a UNESCO World Heritage site in China. These painted artifacts were created during the Qin Shi Huang Di dynasty (first emperor of China). Within 15 seconds of the terra-cotta pieces being unearthed during excavations, the lacquer beneath the paint begins to curl after being exposed to the dry Xian air. The paint would subsequently flake off in about four minutes. The German Bavarian State Conservation Office developed a PEG preservative that when immediately applied to unearthed artifacts has aided in preserving the colors painted on the pieces of clay soldiers.

- PEG is often used (as an internal calibration compound) in mass spectrometry experiments, with its characteristic fragmentation pattern allowing accurate and reproducible tuning.

- PEG derivatives, such as narrow range ethoxylates, are used as surfactants.

- PEG has been used as the hydrophilic block of amphiphilic block copolymers used to create some polymersomes.

Biological Uses

- PEG is commonly used as a precipitant for plasmid DNA isolation and protein crystallization. X-ray diffraction of protein crystals can reveal the atomic structure of the proteins.

- PEG is used to fuse two different types of cells, most often B-cells and myelomas in order to create hybridomas.

- Polymer segments derived from PEG polyols impart flexibility to polyurethanes for applications such as elastomeric fibers (spandex) and foam cushions.

- In microbiology, PEG precipitation is used to concentrate viruses. PEG is also used to induce complete fusion (mixing of both inner and outer leaflets) in liposomes reconstituted *in vitro*.

- Gene therapy vectors (such as viruses) can be PEG-coated to shield them from inactivation by the immune system and to de-target them from organs where they may build up and have a toxic effect. The size of the PEG polymer has been shown to be important, with larger polymers achieving the best immune protection.

- PEG is a component of stable nucleic acid lipid particles (SNALPs) used to package siRNA for use *in vivo*.

- In blood banking, PEG is used as a potentiator to enhance detection of antigens and antibodies.

- When working with phenol in a laboratory situation, PEG 300 can be used on phenol skin burns to deactivate any residual phenol.

Commercial Uses

- PEG is the basis of many skin creams (as *cetomacrogol*) and personal lubricants (frequently combined with glycerin).

- PEG is used in a number of toothpastes as a dispersant. In this application, it binds water and helps keep xanthan gum uniformly distributed throughout the toothpaste.

- PEG is also under investigation for use in body armor, and in tattoos to monitor diabetes.

- In low-molecular-weight formulations (i.e. PEG 400), it is used in Hewlett-Packard designjet printers as an ink solvent and lubricant for the print heads.

- PEG is also one of the main ingredients in paintball fills, because of its thickness and flexibility. However, as early as 2006, some Paintball manufacturers began substituting cheaper oil-based alternatives for PEG.

- PEG is also used as an anti-foaming agent in food – its INS number is 1521 or E1521 in the EU.

Industrial Uses

- Nitrate ester-plasticized polyethylene glycol is used in Trident II submarine-launched ballistic missile solid rocket fuel.

- Dimethyl ethers of PEG are the key ingredient of Selexol, a solvent used by coal-burning, integrated gasification combined cycle (IGCC) power plants to remove carbon dioxide and hydrogen sulfide from the gas waste stream.

- PEG has been used as the gate insulator in an electric double-layer transistor to induce superconductivity in an insulator.

- PEG is also used as a polymer host for solid polymer electrolytes. Although not yet in commercial production, many groups around the globe are engaged in research on solid polymer electrolytes involving PEG, with the aim of improving their properties, and in permitting their use in batteries, electro-chromic display systems, and other products in the future.

- PEG is injected into industrial processes to reduce foaming in separation equipment.

- PEG is used as a binder in the preparation of technical ceramics.

Health Effects

PEG is generally considered biologically inert and safe. However, a minority of people are allergic to it. Allergy to PEG is usually discovered after a person has been diagnosed with an allergy to an increasing number of seemingly unrelated products, including processed foods, cosmetics, drugs, and other substances that contain PEG or were manufactured with PEG.

Factors Influencing PEGylation

Mode of Attachment

The first important aspect in PEGylation is that the PEG coating should be able to withstand the shear forces that can be experienced by the carrier as it moves through the circulating system. Therefore, physical adsorption of PEG on the carrier surface might not be the best technique as there is a strong possibility that the coating might be desorbed in the biological system. Hence, introduction of PEG polymeric chains on the surface of the carrier through covalent binding is preferred and widely practiced, especially in the case of polymeric nanocarrier, liposomes and protein carriers. The covalent linking of PEG to different functional groups present in various carriers is simple as the terminal group in PEG is the reactive –OH group that is also amenable to modifications by other molecules so that the terminal group can be tailored to complement the functionality present in the carrier group. The PEG chains containing amine, carboxyl, methoxy, maleimide, N-hydroxysuccinimide etc. are available. The PEG can be introduced as a block with the polymer forming the carrier core. The block of PEG can be introduced at the end or in the middle depending upon whether a diblock or triblock copolymer is desired. In the case of di- and triblock copolymers containing PEG, the PEGylation is referred to as 'end-on' type, as the PEG groups will be present in the backbone of the polymer.Alternately, the PEG chains can be grafted to the main chain of the polymer. The number of PEG chains attached to a single polymer chain can be modified in this technique. This will influence the surface density as well as morphology of the PEG chains. This type of PEGylation is referred to as 'side-on' type, as PEG is present in the side chains and not in the backbone of the polymer. Figure depicts an end-on type and side-on type of PEGylation in polymers.

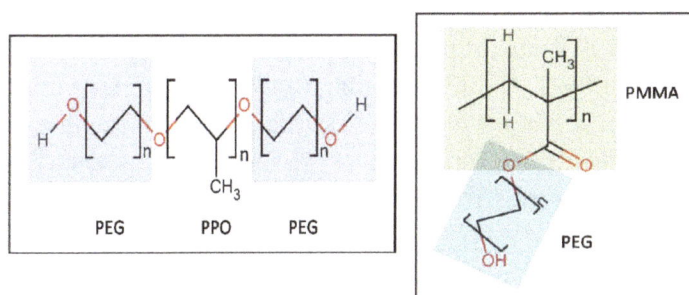

Fig. 3: End-on and side-on type of PEGylation

When a polymer carrier is modified with PEG, due to the highly hydrophilic nature of the PEG chains, they tend to project outwards into the aqueous environment so as to maximize their contact with water. The polymer forming the carrier molecule becomes the core. Experiments have revealed that the PEG segments always try to localize in the surface that is beneficial for the protein repulsion requirement. The nature of PEG linkage determines the configuration of the PEG chains in the carrier. This in turn can influence the protein repelling ability of the PEG chains. Four types of PEG configurations on polymeric carriers have been defined – brush, loop, bunch and star. Figure shows the different types of configurations of PEG on nanocarriers.

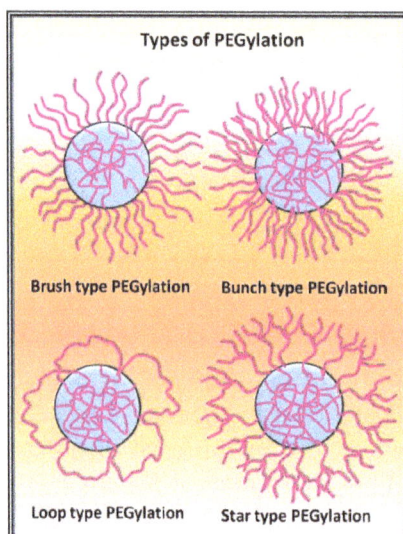

Different types of PEG configurations in nanocarriers

The brush configuration can be achieved with diblock copolymers and is the most commonly employed. The triblock copolymers containing PEG in the central segment form structures with a loop configuration of PEG. The graft polymers containing multiple PEG chains in the side chains form bunch type of configuration. The star configuration is least explored and is obtained using special type of branched copolymer designs. The brush and loop configurations were found to be comparatively more effective than the other configurations in repelling proteins. This difference may mainly arise from the inter-chain steric hindrances that may be posed by the PEG chains in the bunch and star configurations. In other words, the 'end-on' configuration works best for PEGylated carriers!

Molecular Weight of PEG

Another important parameter that can influence the protein repelling characteristics of the PEG coating is the molecular weight of PEG. As PEG is a linear polymer, the molecular weight is directly responsible for its chain length. It has been found that the protein adsorption remains unchanged if the molecular weight of the surface PEG chains is less than 1000 Da. Hence the minimum molecular weight of PEG that needs to be used is 1000 Da. As the molecular weight increases, the protein adsorption decreases up to a molecular weight of 5000 Da, beyond which not much significant decrease in protein adsorption is observed. Increasing the molecular weight beyond 5000 Da will no doubt be able to decrease the opsonization and immune response enabling it to remain in the blood circulation for a longer time. However, it will also retard cell uptake of the nanocarrier as the same factors that retard opsonization and immune response will also contribute to reduction in cell uptake. Hence, it is essential that the molecular weight of the PEG chains be in the range 2000-5000 Da to ensure sufficient retardation of opsonization as well as not compromise too much on the cell uptake of the nanocarrier. Figure represents the influence of the molecular weight of PEG on cell uptake.

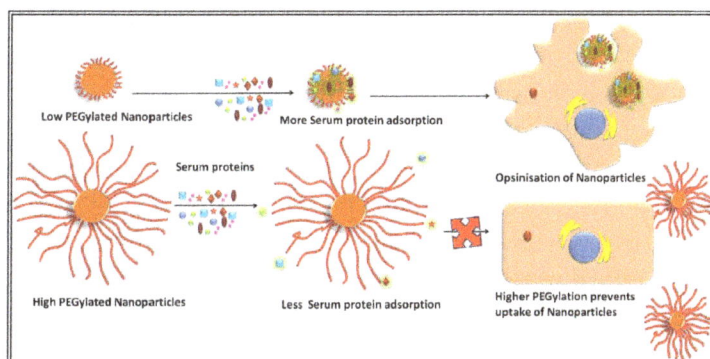

Influence of molecular weight of PEG on cell uptake

Surface Density of PEG Chains

The surface density of the PEG chains is another factor that needs to be in the optimum range for effective stealth characteristics. Ideally, the distance 'd' between the PEG chains on the surface should be 2.2 nm. Higher values of separation will leave gaps on the surface where proteins could adsorb leading to immune recognition.

Attachment of Targeting Ligand

If the surface is totally covered with PEG, then the cell uptake will be reduced considerably. Hence, current approaches involve covalently linking a targeting ligand to the terminus of the PEG chains to ensure that the carrier can selectively enter a target cell using the concept of 'receptor-mediated endocytosis'. The choice of the targeting ligand is based on the expression of surface receptors on the cell surface that are unique to the target cell. To facilitate covalent linking of the targeting ligand, the terminal group of the PEG chain could be suitably chosen.

Why should the targeting ligand be linked to the terminus of the PEG chain and not directly on the surface of the nanocarrier itself? Well, if the targeting ligand were directly present on the surface of the nanocarrier itself, the PEG chains will mask the ligand from its target receptor and hence binding and subsequent entry of the nanocarrier will never occur. On the other hand, if the targeting ligand is on the terminus of the PEG chain, it can selectively bind with the target receptor.

It should be noted that the PEGylation cannot permanently avoid opsonization. The PEGylation only aids in increasing the blood circulation time of the nanocarrier thereby increasing its chances of reaching the target! Thus while PEGylation of a nanocarrier can enhance the residence time of the carrier in blood from a few minutes (less than 10 min in most cases) to a few hours, the ultimate destination of the PEGylated carrier still will be the MPS. Therefore, better strategies to completely inhibit opsonization and MPS activation need to be identified.

Other Alternatives to PEG

Other alternatives to PEGylation have been investigated and a few candidates have been identified to have the potential to impart stealth characteristics to the nanocarrier. Whether these could replace PEG? The answer is still NO, as they have not been demonstrated to be superior to PEG. The list of potential replacements for PEG includes dextran (the soluble form), pullulan, monosialoganglioside (GM1),heparin, poly(vinyl pyrrolidone), poly(acryl amide), poly(2-methyl2-oxazoline) and poly(2-ethyl oxazoline). Among these, dextran has been the most promising. Interestingly, the loop configuration for dextran is not effective in protein repulsion unlike PEG. In the case of dextran, a side- on type of linkage provides better results with the brush configuration and bunch configuration being more effective.The linking of heparin to a nanocarrier offers an additional advantage in that it does not allow platelet adhesion or aggregation on the surface. It also does not allow uptake of the nanocarriers by the macrophages and other members of the MPS. As the heparin chains on the surface also retard opsonization due to steric repulsion effects, its anti-complement, anti-MPS and anti-coagulant action makes it a better option than other candidates being explored for imparting stealth characteristics!

References

- Alan D. Michelson (26 October 2006). Platelets. Academic Press. pp. 3–5. ISBN 978-0-12-369367-9. Retrieved 18 October 2012

- Sigel H, Sigel A, eds. (2009). Metallothioneins and Related Chelators (Metal Ions in Life Sciences). Metal Ions in Life Sciences. 5. Cambridge, England: Royal Society of Chemistry. ISBN 1-84755-899-2

- Nutrition, Center for Food Safety and Applied. "Potential Contaminants - 1,4-Dioxane A Manufacturing By-product". www.fda.gov. Retrieved 2017-05-26

- Rossi, J.J. (2006). "RNAi therapeutics: SNALPing siRNAs in vivo". Gene Therapy. 13 (7): 583–584. PMID 17526070. doi:10.1038/sj.gt.3302661

- Schmaier, Alvin H.; Lazarus, Hillard M. (2011). Concise guide to hematology. Chichester, West Sussex, UK: Wiley-Blackwell. p. 91. ISBN 978-1-4051-9666-6

- Margoshes M, Vallee BL (1957). "A cadmium protein from equine kidney cortex". Journal of the American Chemical Society. 79 (17): 4813–4814. doi:10.1021/ja01574a064

- Iverson, Cheryl, et al. (eds) (2007). "12.1.1 Use of Italics". AMA Manual of Style (10th ed.). Oxford, Oxfordshire: Oxford University Press. ISBN 978-0-19-517633-9

- Basu, A. (2010). "Cellular Responses to Cisplatin-Induced DNA Damage". Journal of Nucleic Acids. 2010: 1–16. doi:10.4061/2010/201367

- American Psychological Association (2010), "4.21 Use of Italics", The Publication Manual of the American Psychological Association (6th ed.), Washington, DC, USA: APA, ISBN 978-1-4338-0562-2

- "Current EU approved additives and their E Numbers". UK Government – Food Standards Agency. Retrieved 21 October 2010

- Fee, Conan J.; Damodaran, Vinod B. (2012). "Biopharmaceutical Production Technology": 199. doi:10.1002/9783527653096.ch7. ISBN 9783527653096

- Vignais, Paulette M.; Pierre Vignais (2010). Discovering Life, Manufacturing Life: How the experimental method shaped life sciences. Berlin: Springer. ISBN 90-481-3766-7

- Yu M, Zhao J, Feng S. Vitamin E TPGS prodrug micelles for hydrophilic drug delivery with neuroprotective effects. International Journal of Pharmaceutics. 2012;438(1-2):98-106

- Kahovec, J.; Fox, R. B.; Hatada, K. (2002). "Nomenclature of regular single-strand organic polymers". Pure and Applied Chemistry. 74 (10): 1921–1956. doi:10.1351/pac200274101921

- Elzoghby A, Samy W, Elgindy N. Protein-based nanocarriers as promising drug and gene delivery systems. Journal of Controlled Release. 2012;161(1):38-49

- Veronese, F. M. (2001). "Peptide and protein PEGylation: A review of problems and solutions". Biomaterials. 22 (5): 405–17. doi:10.1016/s0142-9612(00)00193-9. PMID 11214751

- Rothman, S. S. (2002). Lessons from the living cell: the culture of science and the limits of reductionism. New York: McGraw-Hill. ISBN 0-07-137820-0

- Veronese, FM; Harris, JM (2002). "Introduction and overview of peptide and protein pegylation". Advanced drug delivery reviews. 54 (4): 453–6. doi:10.1016/S0169-409X(02)00020-0. PMID 12052707

- Barret, James (1980). Basic Immunology and its Medical Application (2 ed.). St.Louis: The C.V. Mosby Company. ISBN 0-8016-0495-8

- Tosi, Michael F. (2005-08-01). "Innate immune responses to infection". Journal of Allergy and Clinical Immunology. 116 (2): 241–249. doi:10.1016/j.jaci.2005.05.036

- David Lillicrap; Nigel Key; Michael Makris; Denise O'Shaughnessy (2009). Practical Hemostasis and Thrombosis. Wiley-Blackwell. pp. 1–5. ISBN 1-4051-8460-4

- Andersen, F. A. (1999). "Special Report: Reproductive and Developmental Toxicity of Ethylene Glycol and Its Ethers". International Journal of Toxicology. 18 (3): 53–10. doi:10.1177/109158189901800208

Permissions

We would like to thank the editorial team for lending their expertise to make the book truly unique. They have played a crucial role in the development of this book. Without their invaluable contributions this book wouldn't have been possible. They have made vital efforts to compile up to date information on the varied aspects of this subject to make this book a valuable addition to the collection of many professionals and students.

This book was conceptualized with the vision of imparting up-to-date and integrated information in this field. To ensure the same, a matchless editorial board was set up. Every individual on the board went through rigorous rounds of assessment to prove their worth. After which they invested a large part of their time researching and compiling the most relevant data for our readers.

The editorial board has been involved in producing this book since its inception. They have spent rigorous hours researching and exploring the diverse topics which have resulted in the successful publishing of this book. They have passed on their knowledge of decades through this book. To expedite this challenging task, the publisher supported the team at every step. A small team of assistant editors was also appointed to further simplify the editing procedure and attain best results for the readers.

Apart from the editorial board, the designing team has also invested a significant amount of their time in understanding the subject and creating the most relevant covers. They scrutinized every image to scout for the most suitable representation of the subject and create an appropriate cover for the book.

The publishing team has been an ardent support to the editorial, designing and production team. Their endless efforts to recruit the best for this project, has resulted in the accomplishment of this book. They are a veteran in the field of academics and their pool of knowledge is as vast as their experience in printing. Their expertise and guidance has proved useful at every step. Their uncompromising quality standards have made this book an exceptional effort. Their encouragement from time to time has been an inspiration for everyone.

The publisher and the editorial board hope that this book will prove to be a valuable piece of knowledge for students, practitioners and scholars across the globe.

Index

www.ingramcontent.com/pod-product-compliance
Lightning Source LLC
Chambersburg PA
CBHW061246190326
41458CB00011B/3594